Developing Senior Navy Leaders

Requirements for Flag Officer Expertise Today and in the Future

Lawrence M. Hanser, Louis W. Miller,
Herbert J. Shukiar, Bruce Newsome

Prepared for the Chief of Naval Personnel

Approved for public release; distribution unlimited

RAND NATIONAL DEFENSE RESEARCH INSTITUTE

This research was sponsored by the Chief of Naval Personnel (CNP) and conducted within the Forces and Resources Center of the RAND National Defense Research Institute, a federally funded research and development center sponsored by the Office of the Secretary of Defense, the Joint Staff, the Unified Combatant Commands, the Department of the Navy, the Marine Corps, the defense agencies, and the defense Intelligence Community under Contract W74V8H-06-C-0002.

Library of Congress Cataloging-in-Publication Data

Developing senior Navy leaders : requirements for flag officer expertise today and in the future / Lawrence M. Hanser ... [et al.].

p. cm.

Includes bibliographical references.

ISBN 978-0-8330-4294-1 (pbk. : alk. paper)

1. United States. Navy—Officers—Training of. 2. United States. Navy—Personnel management. 3. Admirals—United States. 4. Leadership. I. Hanser, Lawrence M.

VB203.D48 2008

359.3'310973—dc22

2008019155

Published 2008 by the RAND Corporation

1776 Main Street, P.O. Box 2138, Santa Monica, CA 90407-2138

1200 South Hayes Street, Arlington, VA 22202-5050

4570 Fifth Avenue, Suite 600, Pittsburgh, PA 15213-2665

RAND URL: http://www.rand.org/

To order RAND documents or to obtain additional information, contact

Distribution Services: Telephone: (310) 451-7002;

Fax: (310) 451-6915; Email: order@rand.org

Preface

The array of expertise required to be a successful leader in the U.S. Navy has become more complex. To be a successful Navy leader, it is no longer sufficient to be skilled only at surface, submarine, or air warfare. Additional kinds of expertise are needed to lead and manage the Navy of today and the Navy of the future. Furthermore, like its sister services, the Navy also has a large and distinct core of senior civilian leaders that continues to provide a broad array of in-depth business skills, as well as the continuity and stability of senior leadership.

Navy leaders have become increasingly concerned that senior officers need additional kinds of expertise, besides those traditionally developed in naval officers, to be successful in commanding, leading, and managing the Navy enterprise. In 2002, the Chief of Naval Operations (CNO) established the Office of the Executive Learning Officer (ELO) to lead the Navy in creating learning and experiential opportunities for senior naval leaders to develop these additional kinds of expertise.

Working with the ELO, RAND focused on identifying the expertise requirements of Navy flag billets and joint billets filled by Navy flag officers as a means of understanding what learning and experiential opportunities are needed. The study's results will be of greatest interest to senior leaders in the Navy and those individuals and organizations engaged in the development of naval officers.

This research was sponsored by the Chief of Naval Personnel and conducted within the Forces and Resources Center of the RAND National Defense Research Institute, a federally funded research and development center sponsored by the Office of the Secretary of Defense, the Joint Staff, the Unified Combatant Commands, the Department of the Navy, the Marine Corps, the defense agencies, and the defense Intelligence Community.

For more information on RAND's Forces and Resources Policy Center, contact the Director, James Hosek. He can be reached by email at James_Hosek@rand.org; by phone at 310-393-0411, extension 7183; or by mail at the RAND Corporation, 1776 Main Street, Santa Monica, California 90407-2138. More information about RAND is available at http://www.rand.org.

Contents

Preface iii
Figures ix
Tables xi
Summary xiii
Acknowledgments xxi
Abbreviations xxiii

CHAPTER ONE
Introduction 1
Organization of This Monograph 3

CHAPTER TWO
Background 5
Flag Officers Serve in a Wide Range of Billets 6
What Kinds of Officers Lead the Navy? 8

CHAPTER THREE
A Framework for Understanding Flag Officer Billet Requirements 19
Domain Expertise 20
Leadership, Management, and Enterprise Expertise 23
Identifying the Requirements for Expertise 24
The Job Book Documents Area-of-Expertise and Other Billet Requirements 25

CHAPTER FOUR
A Closer Look at Expertise Requirements 27
Flexibility in Primary Area of Expertise.... 27
Secondary Areas of Expertise.... 29
Importance of Leadership, Management, and Enterprise Expertise.... 30
Relating Expertise Requirements to Pre-Flag Officer Development.... 34
Flag Officer Requirements Have Critical Bottlenecks.... 36
Reconciling the Constraining Requirements.... 41

CHAPTER FIVE
Matching Domain Expertise to Billets in the Navy Flag Officer Force.... 43
Modeling the Average Flows of Inventory in the Flag Officer Ranks.... 44
Application of the Data in the Model.... 47
Deriving Sets of Areas of Expertise from the Billet Database.... 50
Average Times in Grade.... 50
Upper Bounds on Promotion Probabilities and Thresholds on Promotion Rates.... 50
Some Results from a Sample Run of the Model.... 51
Exploring Annual Promotion Rates.... 53
Concluding Remarks.... 58

CHAPTER SIX
RDML Selectees: Comparison with Model-Determined Requirements.... 61
Characterizing the Supply.... 61
Can All the Identified Expertise Pairs Be Used?.... 62
Are All Necessary Expertise Pairs Available Among the Six Cohorts?.... 67
A Final Word on the Gap Analysis.... 67

CHAPTER SEVEN
An Exploration of Future Requirements.... 69
How We Derived Future Expertise Requirements.... 70
Summary of Future Significant Impacts on Required Areas of Expertise.... 71
Organizational and Structural Changes.... 73
Ship-Related Force Structure.... 73

Information Warfare 74
Space Warfare 75
Intel and Foreign Area Officers 76
Terrorism and Counterterrorism 77
Ballistic Missile Defense 77
Anti-Submarine Warfare 78
Special Warfare, Expeditionary Warfare, and Littoral Warfare 78
Stability Operations 79
Readiness, Logistics, Sea Basing, and Training 80
Development Changes 80
Shifts in Advanced Education 80
Shifts in Training: High-Tech Ships with Smaller Crews 81
Operational Strategy Changes 82
Technology Acquisitions 85
Information Technology Acquisitions 86
Submarine Warfare Acquisitions 87
Surface Warfare Acquisitions 90
Logistics and Readiness 91
Expeditionary Warfare Acquisitions 91
Counter-Mine Warfare Acquisitions 94
Special Warfare Acquisitions 95
Stability Operations: No Acquisitions Directly Relate 95
The Relevance of Current Expertise for the Future 95
Potentially Uncaptured Areas of Expertise 96

CHAPTER EIGHT
Conclusions and Recommendations 99
Conclusions 99
Recommendations 102

APPENDIXES
A. Flag Billet Titles 105
B. Definitions of Domain Expertise 119
C. Cross-Functional Expertise 125
D. Survey Screenshots and Additional Definitions Used in the Survey 129

E. Formulation of the Mathematical Program to Determine Average Flows of Inventory Through the Flag Officer Ranks 141

References 149

Figures

2.1. Department of the Navy 7
2.2. Navy Headquarters 8
2.3. Navy Operating Forces 9
2.4. Navy Shore Organization 10
4.1. Flexibility in Primary Expertise Requirements 28
4.2. Most Prevalent Primary Domain Expertise Requirements 29
4.3. Most Prevalent Secondary Domain Expertise Requirements 31
5.1. Comparison of Primary Areas of Expertise in the Data and O-7 Promotions in the Sample Solution 57
5.2. Comparison of Secondary Areas of Expertise in the Data and O-7 Promotions in the Sample Solution 59
D.1. Survey Screen 1: Welcome Page 132
D.2. Survey Screen 2: Introduction 133
D.3. Survey Screen 3: Domain Knowledge Requirements 134
D.4. Survey Screen 4: Enterprise Knowledge, Leadership, and Management Skill Requirements (1 of 2) 135
D.5. Survey Screen 5: Enterprise Knowledge, Leadership, and Management Skill Requirements (2 of 2) 136
D.6. Survey Screen 6: Education Requirements 137
D.7. Survey Screen 7: Foreign Language Requirements 138
D.8. Survey Screen 8: Military/Civilian Fill Requirements 139
D.9. Survey Screen 9: Comments 140
E.1. Flows of Expertise to Billets 143
E.2. Flows of Inventory Across Successive Grades 144

Tables

2.1. Navy Officer Communities and Related Designator Codes.... 12
2.2. Number of Active Flag Officers by Designator 14
2.3. Distribution of Active Navy Captains and Flag Officers by Designator.. 16
3.1. Areas of Domain Expertise... 21
4.1. Proportion of Billets Identifying Specific Areas of Leadership, Management, and Enterprise Expertise as Critical.. 32
4.2. Flag Officer Billets (O-8 and Above) Restricted to One or Two Primary and Secondary Areas of Expertise........ 37
4.3. Billets Calling for Submarine Warfare/Nuclear Propulsion or Submarine Warfare/Ship Engineering and Repair 39
5.1. Distribution of Billets Across Grades............................ 47
5.2. Expertise Used in the Model 48
5.3. Filling the Commander, Naval Network Warfare Command, Billet 52
5.4. O-7 Annual Promotions 54
6.1. Area-of-Expertise Pairs Seen in Five Years of Promotions to RDML... 63
6.2. Primary/Secondary Pairs of Expertise with Supply/Demand or Supply-Only Indications 64
6.3. Missing Primary/Secondary Expertise Pairs...................... 67

Summary

The career paths of Navy officers leave little opportunity for developing depth in an area of expertise outside of what, for most, is their career-long officer designator code. This is especially true for unrestricted line officers. Yet these officers are called on at the pinnacle of their careers as flag officers to lead and manage large Navy enterprises, such as the Naval Sea Systems Command, which engineers, builds, and supports the Navy's ships and combat systems, has nearly 37,000 personnel, and alone accounts for almost a fifth of the entire Navy budget. Former Chief of Naval Operations, Admiral Vernon E. Clark, recognized that the Navy might not be developing the business acumen in its military personnel that is needed to command such organizations and established the position of Executive Learning Officer to address this concern.

Is there a gap in officer development that manifests itself in the flag officer ranks, and if so, what is the nature and size of it? Working with Navy colleagues through the ELO, RAND engaged in a four-step process to address ADM Clark's concerns. First, we determined the kinds of expertise required for successful performance in flag billets. Second, we created a model to identify the kinds of expertise incoming O-7s (officers with a pay grade of Rear Admiral, Lower Half) must have to satisfy flag billet requirements. Third, we compared these model-determined requirements against several years of O-7 selectees, looking for differences in areas of expertise those selectees possessed and the model-determined requirements. Finally, we used Navy planning documents as the basis for forecasting how specific areas of expertise might increase or decrease in importance over the next decade.

The areas of expertise developed in the first step split into two distinct categories: domain-specific areas of expertise and broader, non-domain-specific areas. The domain-specific areas of expertise include many that map well to billet and officer designator codes, such as "surface warfare officer"; some that are currently used as additional qualification designation (AQD) codes, such as "financial management"; and other areas of domain-specific expertise, such as "installation management."

One might reasonably expect that these domain-specific areas of expertise would be required by some specific billets but not by others. For example, a submarine group commander billet requires a submarine warfare officer (112X) and will not be filled by a surface warfare officer (111X). However, the list of areas of expertise also includes expertise in leadership, management, and enterprise perspective. Unlike domain expertise, these kinds of expertise are not domain-specific. Further, they are widely required by most if not all flag officer billets. For example, our data, which are based on surveys of Navy flag officers and are designed to identify the critical areas of expertise to ensure successful performance in their billets, show that expertise in "exercising responsibility, good judgment, authority, and accountability" is critical for virtually all flag officer billets.

We surveyed all Navy flag officers to identify the areas of expertise they consider critical to success in their billets. Because the leadership, management, and enterprise-knowledge areas of expertise are required of most billets, i.e., are nondiscriminatory in terms of informing billet assignment priorities, in step 2, we constructed a model to illuminate the demand for domain-specific expertise in the O-7 flag officer entry cohort. In step 3, we examined the career experiences of six years of O-7 selectees to understand the areas of domain expertise that naval officers bring to the flag officer ranks, in addition to the expertise identified by their officer designator codes. We compared the demand for domain expertise with the supply of domain expertise to identify the magnitude and nature of the development gap. Finally, in step 4, we examined future planned changes to Navy organizations and equipment procurements to identify where the need for such expertise may shift in the future.

Requirements for Leadership, Management, and Enterprise Expertise

We identified 38 specific areas of expertise in leadership, management, and enterprise perspective. Confirming our expectation that these areas of expertise are critical for successful performance in a wide array of flag billets, each of 22 areas of expertise was rated as critical for at least 70 percent of flag officer billets (see Chapter Four, Table 4.1). The most widely cited as critical (for over 90 percent of billets) were

- exercising responsibility, good judgment, authority, and accountability
- motivating, inspiring, and mentoring military personnel
- exercising good judgment, perception, adaptiveness, and common sense to integrate priorities and eliminate irrelevant information
- guiding expectations, managing risk, and achieving results
- resolving conflict and confrontation with and among superiors, peers, and subordinates in a peacetime environment
- influencing and negotiating with people at all levels.

Requirements for Domain-Specific Expertise

Another kind of expertise required in flag billets is "domain-specific." By this we mean expertise related to operational or functional areas, such as surface warfare or acquisition, as distinct from the leadership, management, and enterprise expertise requirements of flag billets. We characterize requirements for domain-specific expertise as "primary" or "secondary." By this we simply mean to answer the question, "What is the most critical area of domain expertise required in a billet?" and "What is the second-most critical area of domain expertise required in a billet?" For many flag billets, we find that several different primary areas of expertise are substitutable. For example, for a number of flag billets, such as carrier battle group command billets, it is critical to have a warfare expertise, but it may be in any one of surface warfare, aviation warfare, or submarine warfare. Some billets are inflexible in primary domain-specific expertise requirements: Approximately 30 percent of billets require one specific primary area of domain expertise. On the other hand, for almost 10 percent of flag billets, a flag officer with any primary area of domain expertise could perform successfully (see Chapter Four, Figure 4.2).

Many more areas of domain-specific expertise are indicated as critical for flag billets than a single individual could possibly develop in any depth over the course of a Navy career. We interpret this to mean that flag officers must be well-informed consumers of information in multiple domains. Although they can perhaps develop depth in two or three domain areas, they will need to rely on staff for depth of domain expertise they lack. Perhaps because it is recognized that flag officer billets are accompanied by a staff, our data show much more flexibility in the requirements for secondary areas of domain-specific expertise in flag billets. The most widely cited critical secondary domain-specific areas of expertise are financial management, joint and combined warfare, strategic plans and policy, warfare resources management, and information warfare (see Chapter Four, Figure 4.3).

The Job Book Documents Each Billet's Areas of Expertise

The domain-specific and leadership/management/enterprise areas of expertise associated with each flag billet are available electronically through a job book that is maintained and periodically refreshed by the ELO. In addition to the areas of expertise, the job book's entry for each flag billet also indicates if the billet has special education requirements.

In the initial flag officer survey to identify the flag billets' requirements, each billet had at least two survey responses: one from the current incumbent and one from the preceding incumbent. After reconciliation and senior flag review of these responses, the *synthesized* requirements were placed in the job book. Today, the ELO has responsibility for refreshing each billet's entry in the job book by querying the new incumbent, after he's been in the job for six months, to review and update the billet's requirements.

Is There a Gap in Officer Development?

The gap analysis was performed by examining the area of expertise characteristics of the 2001–2006 O-7 selectee cohorts. There was no way for us to judge whether individual flag officers in these cohorts had the leadership, management, and enterprise perspectives indicated above as critical, so no gap analysis for those areas of expertise was possible. We did conduct a gap analysis for the domain-specific areas of expertise in those cohorts.

Although there is a great deal of flexibility in meeting today's requirements for primary and secondary domain-specific expertise in flag billets, the results of our modeling (discussed in Chapter Five) suggest that, to meet today's flag officer billet requirements, the Navy would especially benefit from developing the following nine pairs of primary and secondary domain-specific areas of expertise in its officers:

- Air Warfare/Joint and Combined Warfare
- Air, Surface, or Submarine Warfare/Public Affairs
- Surface Warfare/Logistics and Readiness
- Surface Warfare/Strategic Plans and Policy
- Submarine Warfare/Ship Engineering and Repair
- **Special Warfare/Counterterrorism**
- Supply Management/Financial Management
- **Civil Engineering/Financial Management**
- **Intelligence/Joint and Combined Warfare**

Many of the pairs of domain-specific expertise shown in our model solution (see Chapter Five, Table 5.4) were held by members of the 2001–2006 O-7 entry cohorts (see Chapter Six, Table 6.2). However, no flag officers in these cohorts had the requisite expertise to match three pairs of expertise shown in bold above (see Chapter Six, Table 6.3). It is possible that some individuals did have these pairs of expertise, but it was not evident to us from the information we had available to make such judgments.

Further, our gap analysis focused only on Rear Admiral, Lower Half (RDML) selectees. A more comprehensive analysis gap analysis that focuses on the O-6 (Captain) population viewed as competitive for flag rank would be useful. The presence in this population of officers with the three primary/secondary pairs found missing in the RDML selectees would suggest that gaps could be addressed through RDML promotion board precepts. However, the absence of these competency pairs from the competitive O-6 population would suggest more serious officer development issues.

How Might Future Changes in the Navy Affect Requirements for Expertise?

We examined the Navy's structure, its force development, its doctrinal documents, and its technology acquisitions for the past decade and the next decade to forecast how the demand for domain-specific expertise may change in the future. The areas of domain-specific expertise with the strongest evidence of increasing future importance to the Navy are (see Chapter Seven):

- Information Warfare
- Information Operations
- Information Technology
- Surface Warfare
- Submarine Warfare
- Special Warfare
- Expeditionary Warfare
- Intelligence
- Logistics and Readiness
- Anti-Submarine Warfare
- Littoral Warfare
- Sea Basing.

Recommendations

If flag officer development is to keep pace with the changing demands for expertise in flag billets, it is important for the Navy to maintain an up-to-date database of requirements. A nascent mechanism for doing this exists in the ELO but will need to be nurtured and supported if it is to survive. ELO is particularly well suited to this task because of its close connection to flag officer development.

Some areas of expertise uncovered by our research as critical for success in Navy billets are not yet well defined. For example, many flag officers identified expertise in Financial Management as critical. However, face-to-face interviews with senior flag officers, conducted after the surveys were completed and reviewed, suggest that Financial Management was a shorthand way for them to convey that it is broader expertise in business-related topics that is critical for performance. We believe the same could be said about other areas of expertise identified in this research, such as Warfare Resources Management and Joint and Combined Warfare, but we have not conducted the face-to-face interviews needed to illuminate this. Fleshing out the details associated with these topics, either through face-to-face interviews or other mechanisms is necessary for the Navy to understand the nature of the developmental opportunities its flag officers require to meet these requirements.

Finally, while the gap analysis does suggest the need to develop a number of primary/secondary domain expertise pairs among pre-flag officers who are deemed to be competitive for flag selection, that analysis also demonstrates that the Navy is for the most part doing a good job in providing pre-flag officers with the necessary domain expertise characteristics to serve effectively in flag billets.

Acknowledgments

We would like to extend our warmest thanks and appreciation to VADM Phil Quast (USN ret.), the Navy's Executive Learning Officer, and to his brother, RADM Harry Quast (USN ret.), for their support and assistance throughout this project. In addition, several members of the ELO staff assisted with gathering and compiling data, and with the untold administrative details that need attention when so much contact with flag officers is required. These staff included Jeff Munks, Tony Lucas, James Sanderson, Roxane Lynn, Cindy Kohatsu, and Cecilia Lucas. Lastly, we wish to express our admiration and thanks to the many flag officers who devoted time from their busy schedules to respond to the two iterations of the survey that were fielded and to the many hours of conversation with us that helped us to understand the expertise requirements of Navy flag officer billets. Special thanks go to VADM Gerry Hoewing (USN ret.) for devoting so much of his time as Chief of Naval Personnel to critiquing and encouraging our work and for reviewing all survey responses in order to review and help resolve differences among different surveys of the same or similar billets.

Abbreviations

ADM	Admiral (O-10)
AQD	additional qualification designation
ASW	Anti-Submarine Warfare
BAMS UAV	Broad Area Maritime Surveillance Unmanned Aerial Vehicle
BMD	Ballistic Missile Defense
CEC	Cooperative Engagement Capability
CG(X)	Navy's Next Generation Cruiser
CNO	Chief of Naval Operations
DCNO	Deputy Chief of Naval Operations
DD(X)	Navy's Next Generation Destroyer
DoD	Department of Defense
ELO	Executive Learning Officer
FAO	Foreign Area Officer
FIWC	Fleet Information Warfare Center
FMO	Flag Matters Office
HSV	high-speed vessel
IO	Information Operations
IT	Information Technology

IW	Information Warfare
LCAC	Landing Craft Air Cushion
LCS	Littoral Combat Ship
NAVSEA	Naval Sea Systems Command
NETWAR	Naval Network Warfare Command
NMCI	Navy/Marine Corps Intranet
NNSOC	Naval Network and Space Operations Command
NPS	Naval Postgraduate School
O-6	Captain
O-7	Rear Admiral, Lower Half (RDML)
O-8	Rear Admiral, Upper Half (RADM)
O-9	Vice Admiral (VADM)
O-10	Admiral (ADM)
OAMCM	Organic Airborne Mine Countermeasure
PEO	Program Executive Officer
RADM	Rear Admiral, Upper Half (O-8)
RDML	Rear Admiral, Lower Half (O-7)
RL	Restricted Line
SSBN	Ballistic Missile Submarine—Nuclear
SSGN	Guided (Cruise) Missile Submarine—Nuclear
URL	Unrestricted Line
USMC	U.S. Marine Corps
USN	U.S. Navy
VADM	Vice Admiral (O-9)

CHAPTER ONE

Introduction

Could U.S. Navy officers be better prepared to become flag officers? In remarks made at a seminar for Navy senior executives in September 2003, then Chief of Naval Operations (CNO), Admiral Vernon E. Clark, noted that senior Navy leaders need business skills if the Navy is to succeed in a competitive environment:

> We set out the Executive Business Course because we were convinced that we were not investing in the executive leadership corps, and on the uniformed side we weren't even close to having flag officers who knew enough about the business to be able to do this right. (Clark, 2003, p. 2)

In light of Admiral Clark's voiced concern, the Navy asked RAND to help it understand the areas of expertise[1] required to meet the demands of flag officer billets and the shortfalls, if any, in current development efforts. By "expertise" we mean the ability of an individual to perform tasks associated with an occupation, sub-occupation, or field. For example, one could have expertise in "surface warfare," "supply," or "strategic planning," to name a few. Our concept of "expertise" also admits of *degrees* or *depth* of expertise in the sense that one individual may have greater expertise in a given area than another indi-

[1] In current usage, one could substitute the term *competency* for our use of the word *expertise*. However, we have chosen to avoid the term *competency* because of the current confusion with regard to its meaning (see Shippmann et al., 2000).

vidual or that one billet may require greater expertise in a given area than another billet.

Flag billets differ significantly in their requirements for occupational expertise, and individuals differ widely in the kinds of occupational expertise they possess. Furthermore, military careers are constantly under pressure to become more narrowly occupationally focused (e.g., consider the career path that a Navy fighter pilot must trod to become a flag officer). Yet we know that flag billets often require depth of expertise in several occupational areas (e.g., air warfare and acquisition) and broad familiarity with a host of others. As a result, much can be gained by understanding the differential requirements for and the supply of occupational expertise. Thus, the primary focus of this research is on occupational expertise.

Why do we touch only lightly on the requirements for expertise in leadership when these ranks represent the pinnacle of Navy military leadership positions? Although expertise in leadership is critical to success at these levels, our data show that it is essentially uniformly critical for all flag ranks and billets. Moreover, individuals who reach flag ranks have already passed numerous tests of their leadership skills, so there is little differentiation in either the demand for or supply of leadership expertise. In short, there is good evidence that leadership development is already well understood and practiced in the Navy.

The goal of this research is to strengthen both the Navy's tactical and strategic perspectives on flag officer development in terms of occupational expertise. Our intention is for the Navy to be able to answer questions such as the following: When a flag officer is given a new assignment, what are the just-in-time development needs for that specific flag officer to fill that specific billet? More strategically, what are the domains in which the Navy should focus officer development for the future?

Answering this latter question is not as straightforward as it may seem, because all flag grades must be filled by officers who enter at O-7 and flow upward, and the profiles of expertise requirements may differ from grade to grade. We chose to estimate the steady-state requirements for expertise in entering O-7 cohorts because it is the entry point to flag grades.

Our analysis confirms Admiral Clark's concern that there are gaps between the kinds of expertise that are required of flag officers and the expertise that flag officers possess. Although few new flag officers possess depth of expertise in a domain outside of their career-long orientation, almost all flag billets have specific requirements for such depth. For example, although the Navy traditionally assigns unrestricted line (URL) officers as regional commanders, the officers themselves recognize that an additional degree of business acumen is necessary for their success.

Briefly, our research included four steps. First, we systematically determined the kinds of expertise that are required for successful performance in flag billets. Second, we created a model to estimate the steady-state demands for expertise in incoming rear admirals that will satisfy the expertise requirements for flag billets at all ranks. Third, we reviewed several years of rear admiral (O-7) selectees and looked for differences between this historical supply of, and the steady-state demand for, expertise in incoming cohorts of rear admirals. Finally, we examined Navy planning documents to project how required flag officer expertise might change in the future.

Organization of This Monograph

The next chapter provides some background about the Navy and its flag officers that suggests how the Navy can improve the way it develops flag officers. Chapter Three describes the framework that we adopted for understanding the kinds of expertise that flag officer billets require and how we use this framework to understand the gaps between the supply of, and demand for, expertise. Chapter Four examines the expertise requirements of billets in greater detail. Chapter Five describes a solution for a mix of entry-level Rear Admiral Lower Half (RDML) expertise to support the demands for expertise in flag officer billets. Chapter Six reports on the kinds of expertise that we identified among flag officers who were selected to RDML and the gaps between the supply of and demand for expertise. Chapter Seven explores the near future of the Navy and the changes in demands for flag officer expertise that may be expected to result, and Chapter Eight describes our conclusions and includes recommendations for Navy flag officer development.

CHAPTER TWO

Background

Although the popularly held image of the Navy is one of warships at sea and fighter aircraft in the air, relatively few members of the Navy are deployed operationally at any given point in time. For example, in July 2006, a total of only 36,498 out of 486,299 Navy personnel (354,703 active duty and 131,596 Ready Reserve) were deployed.[1] If we include civilian personnel, the Navy pays approximately 662,000 salaries, not including contractors. There are two primary reasons why such a small proportion of the Navy payroll is deployed at any time. First, personnel and equipment are routinely rotated back to their homeports for training and maintenance. Second, the majority of personnel are needed to create and maintain the infrastructure that supports deployed naval operations. For example, the Naval Sea Systems Command (NAVSEA) is responsible for engineering, building, and supporting ships and combat systems. Its leaders oversee a budget that accounts for one-fifth of the entire Navy budget and approximately 37,000 military and civilian personnel. Naval Air Systems Command (NAVAIR) is smaller but has similar responsibilities for Navy aviation. Additional Navy shore establishments include Naval Facilities Engineering Command, Naval Supply Systems Command, Space and Naval Warfare Systems Command, Naval Education and Training Command, Naval Meteorology and Oceanography Command, and others.

---

[1] The figures presented in this paragraph are taken from the "Status of the Navy" Web page (U.S. Navy, 2007), accessed July 6, 2006.

Flag Officers Serve in a Wide Range of Billets

The variety of expertise required of Navy flag officers is evident in the organizations they lead or serve in and in the titles of the billets they hold. Navy flag officers can be found in all major organizations of the Navy, as well as many Department of Defense (DoD) and joint and combined military organizations.

There are four major distinct sub-organizations within the Navy that each place different demands for expertise on Navy leaders: the Department of the Navy (i.e., the Navy Secretariat; Figure 2.1), Navy Headquarters (Figure 2.2), Navy Operating Forces (Figure 2.3), and the Navy Shore Organization (Figure 2.4).

As the major repository of civilian leadership of the Navy, political appointees and civilians in the Senior Executive Service hold many key positions in the Department of the Navy. Flag officers also serve in several key positions in the department. For example, a handful of flag officers serve as deputy assistant secretaries. Among the additional key billets that flag officers hold within the Department of the Navy are Judge Advocate General of the Navy, Deputy Judge Advocate General of the Navy, Chief of Legislative Affairs, Chief of Information, and Naval Inspector General.

Flag officers hold virtually all of the top leadership positions within Navy Headquarters (Figure 2.2), Navy Operating Forces (Figure 2.3), and the Navy Shore Organization (Figure 2.4). Within these organizations they are often seconded by members of the Senior Executive Service—approximately 220 civilians who provide stability and additional technical expertise in core leadership billets.

The variety of expertise required of flag officers is also evident from the titles of billets that flag officers hold (see Appendix A). Over 40 percent of billets are labeled as commanders, vice commanders, deputy commanders, or assistant commanders. Over 20 percent of billets are labeled as directors or deputy directors. The remaining billets include those deeply engaged in acquisition programs, such as the various program executive officer (PEO) billets and the heads of educational institutions, including the Commandant of the Naval Academy and the President of the Naval War College. The variety of organiza-

Figure 2.1
Department of the Navy (current as of October 23, 2007)

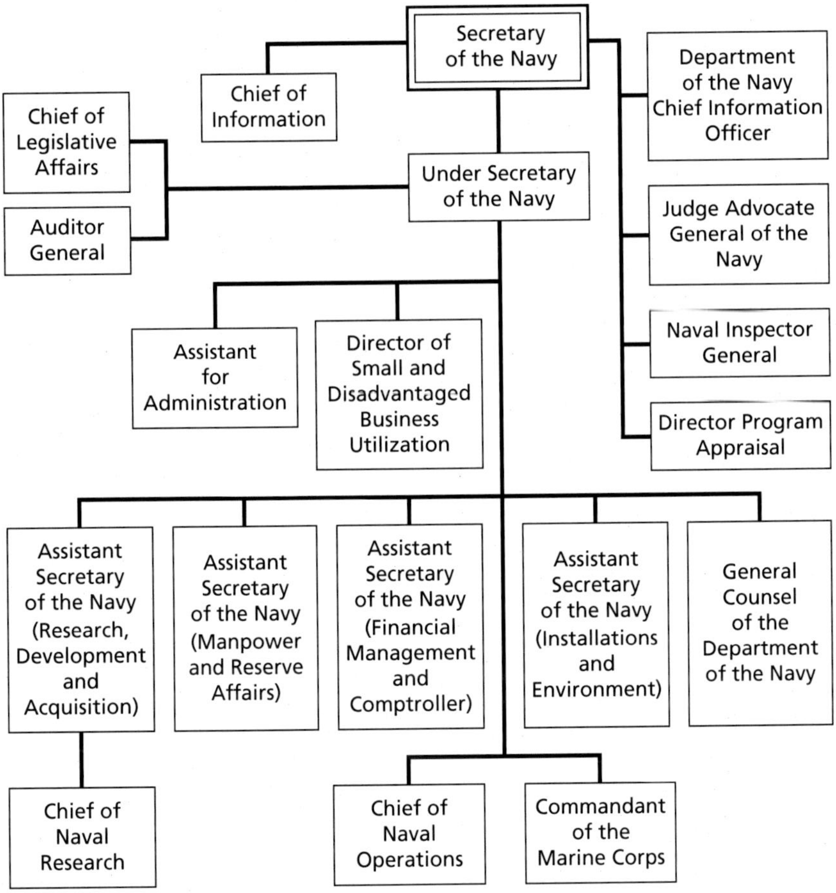

RAND *MG618-2.1*

tions flag officers command and direct includes the Submarine Warfare Division, the White House Military Office, the Fiscal Management Division (N82), Maritime Partnership Programs, Aviation and Aircraft Carrier Plans and Requirements, Navy Expeditionary Combat Command, Navy Recruiting Command, Naval Facilities Engineering Command, and all strike groups.

Figure 2.2
Navy Headquarters (current as of October 23, 2007)

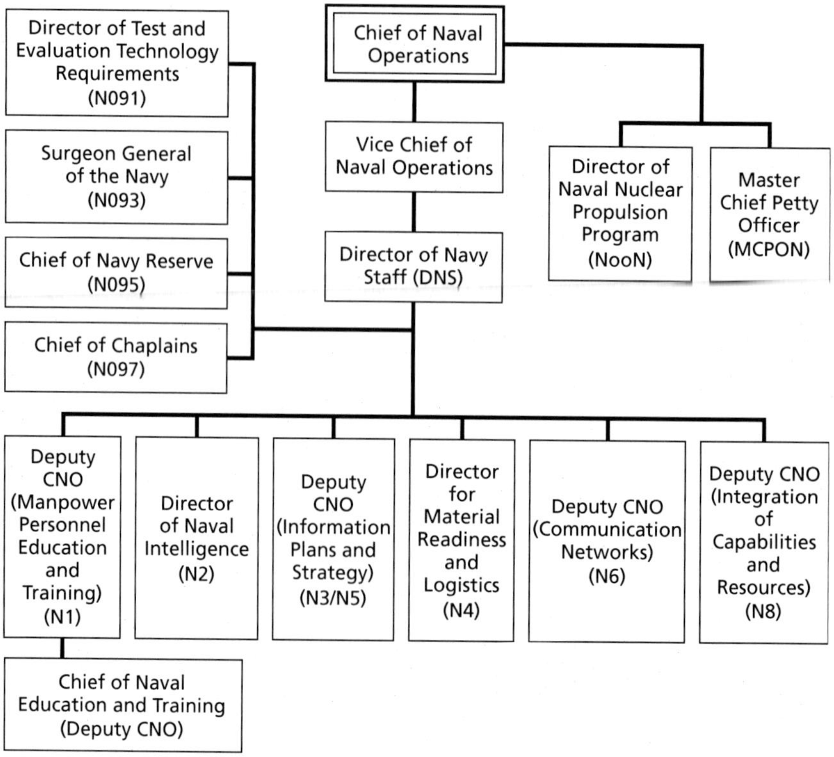

RAND *MG618-2.2*

What Kinds of Officers Lead the Navy?

Like many organizations, the Navy fills the majority of its senior military leadership billets with personnel who represent the core military activities of the organization—unrestricted line officers (i.e., Aviation, Special Operations, Special Warfare, Submarine, and Surface Warfare officers). Using URL officers serves both practical and symbolic purposes. For example, even though the flag billets such as the PEO billets are not operational warfighting billets, the Navy benefits in program execution from the practical fleet experience these individuals have had. The Navy benefits in a different way from filling billets such as the

Figure 2.3
Navy Operating Forces (current as of October 23, 2007)

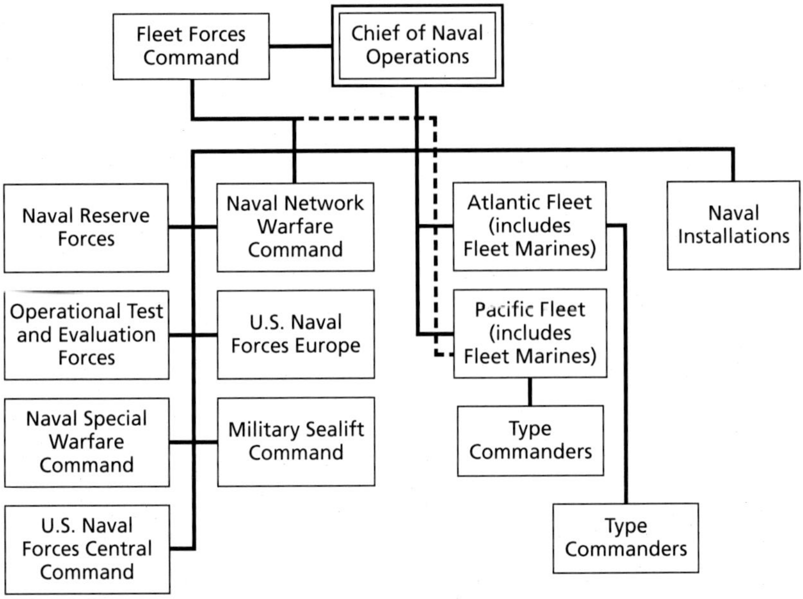

RAND *MG618-2.3*

President of the Naval War College or Chief of Legislative Affairs with URL officers. Relative to their representation among Navy captains, URL officers hold a disproportionate share of flag billets (Table 2.3).

Unrestricted line officers spend their careers learning various aspects of the art of warfare. While learning the art of war—an art that is increasingly fast-moving and technologically complex—there is limited time and opportunity for these officers to develop all of the additional kinds of expertise required to lead large business entities such as NAVSEA. A naval officer's career consists of a series of at-sea and ashore rotations that leave little time for developmental opportunities beyond those focused on his or her primary area of expertise. Much of officer training is focused on developing expertise associated with his or her "designator."[2] For example, the focus of a naval aviator's time

2 According to the *Manual of Navy Officer Manpower and Personnel Classifications, Volume I*, "The officer designator codes are four-digit numbers used to group officers by catego-

Figure 2.4
Navy Shore Organization (current as of October 23, 2007)

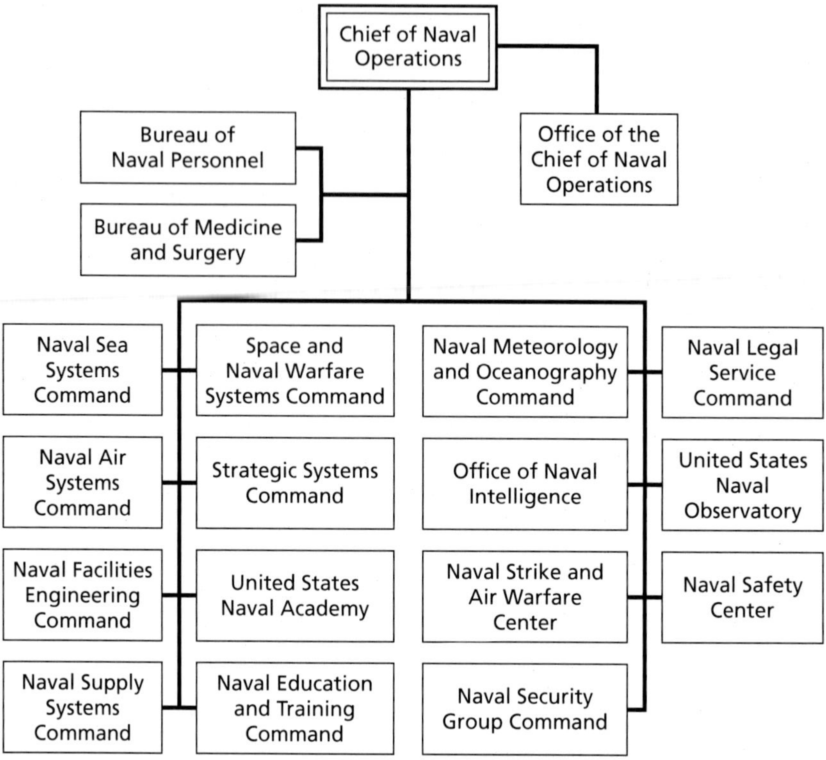

RAND *MG618-2.4*

is spent in the cockpit or on direct support of flying or fleet operations. The designator code for a URL officer "qualified for duty involving flying heavier-than-air, or heavier and lighter-than-air type of aircraft as a pilot" is 131X. Submariners spend substantial time learning how to operate and command nuclear submarines. The relevant designator code for a URL officer who is qualified in submarine warfare is 112X.

---

ries for personnel accounting and administrative purposes and to identify the status of officers. These codes identify, through the first three digits, the categories in which officers are appointed and/or designated and, through the fourth digit, the status of the officers within the various categories" (U.S. Navy, 2006, p. A-3). A complete list of billet and officer designator codes can be found in this reference.

Surface ships, even those more modest in size, such as destroyers, are small cities afloat. Officers and seamen operate equipment that desalinates water and generates electricity, as well as maintain and operate complex communications equipment and advanced weapon systems. As sailors advance in experience, both officers and enlisted personnel serve in billets with increasing leadership responsibilities; for members of the URL, these responsibilities frequently have direct connections to the art of warfare (see Figure 2.3).

This is not to say that a Navy career consists only of developing expertise in different aspects of the art of warfare. Officers serve regular shore tours filling staff billets in the Headquarters, Department of Navy (DON), other Navy and defense organizations, and other government agencies where they undoubtedly develop skills that are relevant to business operations (see Figures 2.2 and 2.4). They also participate in professional military education (PME) programs offered on a wide range of topics, and in civilian education programs. These programs are designed to develop expertise that will be needed in future assignments.

Navy officers trod well-worn career paths that are documented in Navy publications and etched in the minds of young naval officers (Table 2.1 shows the breadth of the varied officer communities[3] that exist in the Navy). These paths have served the Navy well in the past and in many ways will continue to serve the Navy positively into the future. However, the existence of these paths presents a double-edged sword for the Navy, especially in the context of the Navy's legitimate dependence on the URL to provide its core of senior leaders. On one hand, the paths have proven effective in developing focused skills in specific occupational areas. As a result, for example, young aviators are prepared for commanding an air wing, and submariners are prepared for commanding a submarine squadron. On the other hand, individuals are discouraged from straying very far from these beaten paths. If a

[3] A "community" is a group of officers holding one of several related officer designator codes (see Table 2.1). Community managers "are assigned by the CNO to represent the special interests and provide management advice for the respective specialty categories" (U.S. Navy, 2006, p. A-3).

Table 2.1
Navy Officer Communities and Related Designator Codes

Community	Officer Designator Code
Unrestricted line (URL)	
Aviation	13XX
Special Operations	114X
Special Warfare	113X
Submarine	112X
Surface Warfare	111X
Restricted line (RL)	
Aerospace Engineering Duty	151X
Aerospace Maintenance Duty	152X
Information Warfare	161X
Engineering Duty	14XX
Human Resources	120X
Information Professional	160X
Intelligence	163X
Meteorology and Oceanography	180X
Public Affairs	165X
Staff corps	
Chaplain Corps	410X
Civil Engineer Corps	510X
Judge Advocate General's Corps	250X
Medical Corps	210X
Medical Service Corps	230X
Dental Corp	220X
Nurse Corps	290X
Supply Corps	310X
Other	
Acquisition Professional (AP)	AQD codes
Limited Duty Officer/Chief Warrant Officer (LDO/CWO)	

AQD = additional qualification designation.

URL officer strays from the path, it may mean decreased opportunities for command, an important stepping-stone to senior leadership billets. And though many senior leadership billets require the kinds of expertise developed in the URL, many other senior leadership billets may rely on expertise in managing large business organizations, a type of expertise not explicitly developed in the URL. This may be especially important because of the large proportion of flag officers who come from the URL—as of May 2007, approximately 71 percent of Navy flag officers have URL designators (see Table 2.3).

Table 2.2 displays the distribution of active flag officers (people) by designator and rank as of May 2007 and demonstrates the dominance of URL officers in flag ranks. The table is sorted in descending order by the total number of flag officers with a given designator. The most prevalent flag officers are surface warfare–qualified URL officers (53). However, the aviation community, including both pilots and naval flight officers, accounts for 68 flag officers—15 more than the number of surface warfare–qualified flag officers.

Table 2.3 shows how the distribution of officer designators among captains and flag officers compares. Notably, while the unrestricted line represents only 42.5 percent of active-duty captains, 71 percent of the flag officers are in the unrestricted line.

The assignment of flag officers is managed and guided very closely by the CNO and a small staff in the Flag Matters Office (FMO). There are congressionally imposed limits on the absolute number of active-duty flag officers and rules regarding the timing of selections to flag officer ranks and promotions[4] to flag officer ranks. Currently, the Navy is limited to 216 active-duty flag officers. Flag officers who hold certain joint duty assignments or certain positions in unified and specified commands or in some other agencies do not count toward this limit.[5] The process of selecting and assigning flag officers is a chess game of matching individuals to billets to both make the best use of the kinds of expertise each officer has and to prepare flag officers for expected or

---

4 Individuals are "selected," but until there is room in the flag officer ceiling, they are not "promoted."

5 U.S. Code Title 10, Subtitle A, Part II, Chapter 32, §§526–528.

Table 2.2
Number of Active Flag Officers by Designator

Designator Description	Officer Designator Code	ADM	VADM	RADM	RDML	Grand Total
Unrestricted Line Officer who is qualified in Surface Warfare	111X	4	11	15	23	53
Unrestricted Line Officer who is qualified for duty involving flying heavier-than-air, or heavier- and lighter-than-air type of aircraft as a pilot	131X	4	5	19	18	46
Unrestricted Line Officer who is qualified in Submarine Warfare	112X	2	5	9	14	30
Unrestricted Line Officer who is qualified for duty involving flying heavier-than-air or heavier- and lighter-than-air type aircraft as a Naval Flight Officer	132X	1	2	5	14	22
Active-duty Nurse Corps, Medical Service Corps, Medical Corps, or Dental Corps Officer in rank of O-7	2XXX		1	5	9	15
Supply Corps Officer	310X			5	7	12
Engineering Duty Officer who is qualified as a Ship Engineering specialist IAW MILPERSMAN 1210-190	144X		1	2	5	8
Unrestricted Line Officer who is qualified in Special Warfare	113X		2	1	3	6
Civil Engineer Corps Officer	510X		1	2	3	6
Restricted Line AED Flag Officer	150X		1	1	4	6
Special Duty Officer (Intelligence)	163X		1	1	2	4
Special Duty Officer (Information Warfare)	161X			1	2	3
Special Duty Officer of the Information Professional Community who provides expertise in information, command and control, and space systems through the planning, acquisition, operation, maintenance, and security of systems	160X		1	1		2
Judge Advocate General Corps Officer	250X			2		2

Table 2.2—Continued

Designator Description	Officer Designator Code	ADM	VADM	RADM	RDML	Grand Total
Restricted Line Officer of the Human Resources Community	120X			1	1	2
Chaplain Corps Officer	410X			1	1	2
Special Duty Officer (Public Affairs)	165X				2	2
Unrestricted Line Officer who is not qualified in any warfare specialty or in training for any warfare specialty	110X		1			1
Unrestricted Line Officer who is a Special Operations officer by virtue of training in the EOD, DIV/SAL, and EOM functional areas	114X				1	1
Special Duty Officer (Oceanography)	180X				1	1
Grand Total		**9**	**32**	**69**	**110**	**220**

SOURCE: U.S. Naval Register (undated), Data current as of May 11, 2007.

NOTES: IAW MILPERSMAN = in accordance with Military Personnel Manual, AED = aerospace engineering duty, EOD = explosive ordnance disposal, DIV/SAL = diving/salvage, EOM = explosive ordnance management.

Table 2.3
Distribution of Active Navy Captains and Flag Officers by Designator

Officer Designator	Percentage Who Are Captains	Percentage Who Are Flag Officers
Aviation Duty Officer (13XX)	1.7	2.7
Aerospace Engineering Duty Officer (Aerospace Engineering)	0.9	
Aerospace Engineering Duty Officer (Aviation Maintenance)	0.3	
Civil Engineer Corps	2.4	2.7
Chaplain Corps	2.6	0.9
Special Duty Officer (Cryptology)	1.1	1.3
Dental Corps	8.3	
Engineering Duty Officer	3.4	3.6
Special Duty Officer (Human Resources)	1.2	0.9
Special Duty Officer (Intelligence)	2.6	1.8
Special Duty Officer (Information Professional)	1.3	0.9
Judge Advocate General's Corps	2.2	0.9
Limited Duty Officer (Line)	0.9	
Limited Duty Officer (Staff)	0.1	
Medical Corps/Senior Healthcare Executive	12.7	5.8
Medical Service Corps	5.0	0.4
Nurse Corps	4.0	0.4
Special Duty Officer (Public Affairs)	0.8	0.4
Special Duty Officer (Oceanography)	0.6	0.9
Supply Corps	5.6	5.4
Unrestricted Line Officer	42.5	71.0

SOURCE: U.S. Naval Register (undated), data current as of May 9, 2007.

planned follow-on assignments.[6] For example, prior to becoming the Chief of Naval Personnel, a flag officer may serve in several assignments in "personnel" related organizations. Substantial complexity arises from the chain of decisions that a single vacancy creates. For example, when a three-star billet becomes vacant it is typically filled either by reassigning another three-star flag officer or by assigning a two-star flag officer and promoting him or her to three-star rank. In either case, a cascading chain of vacancies occurs, and a plan must be created and executed for filling each of these vacancies at nearly the same time.

The assignment problem is exacerbated and the potential for successful performance in flag billets suffers when officers with the required combinations of expertise are sparse. As noted above, there are limited opportunities for URL officers to develop functional expertise outside of their warfighting areas. As a result, many achieve flag officer rank with substantial experience across the Navy but with limited expertise outside of areas associated with their designator code. When assignment decisions are made under these circumstances, two outcomes necessarily occur. First, the CNO and FMO will have fewer well-qualified individuals to choose from for any given assignment. Second, more billets are likely to be filled suboptimally, with individuals who do not have the requisite combination of expertise for the billet.

Finally, flag officer assignments, like decisions made in many organizations when filling senior-level jobs, are often made for reasons more complex than simply matching expertise requirements. For example, flag officers are often assigned to a billet outside of their primary area of expertise to gain experience in another area to prepare them for a later assignment, or for important symbolic reasons. Obviously, many competing considerations come into play when assigning flag officers to billets.

The next chapter provides a framework for understanding the expertise requirements of Navy flag officer billets.

---

6 This is related to the notion of "using" versus "development" tours expressed in Harrell et al. (2004).

CHAPTER THREE

A Framework for Understanding Flag Officer Billet Requirements

The design of our framework for required expertise was driven by the goals of our research: to define the demand for expertise, to evaluate the supply of expertise, and to articulate targets for development of expertise, all in language that speaks clearly to the Navy. Our experience with other organizations suggests that individuals in senior leadership positions require expertise of several kinds: domain (operational or functional expertise), leadership, management, and enterprise. We define these kinds of expertise in greater detail below.

While a billet may require expertise in a specific operational or functional domain, all flag officer billets require some degree of leadership, management, and enterprise expertise. By itself, no single kind of expertise is sufficient for successful performance in a flag officer billet. As Gabarro (1987) noted, "The all-purpose general manager who can be slotted into just about any organization, function, or industry exists only in management textbooks" (p. 68). Domain-specific expertise is also necessary for successful performance in senior Navy leadership billets.

By the time Navy officers achieve the rank of captain (O-6), they have been carefully screened and selected for evidence of having developed expertise in leadership, some aspects of management, and in the Navy and/or defense enterprise itself. Indeed, assignments in the joint arena are required before an O-6 can be promoted to flag rank in any military service. Such assignments provide important experience in the defense enterprise. However, an individual is unlikely to have developed substantial expertise in more than one operational or functional domain in addition to that represented by his or her designator code. One reason is that there are many different command and noncommand experiences a sailor must have successfully completed within his or her Navy community to have achieved the rank of captain and to be competitive for selection to flag rank. There is precious little time in a Navy career to be spent learning about and gaining expertise in other domains.

Domain Expertise

We define expertise in a domain as the set of content-oriented knowledge, skills, abilities, and other personal characteristics normally associated with an occupation or career field in the Navy. For example, one area of domain expertise is Surface Warfare. Table 3.1 lists the domain areas we used in this research. The list was developed in conjunction with the Navy from accepted lists of domain-specific identifiers in use today, and the areas of domain expertise correspond roughly to officer designator codes, with some additions and modifications. Our definitions for each area of expertise are presented in Appendix B.

For each Navy flag officer billet, we identify the critical kinds of domain expertise that are absolutely essential to performance. Our benchmark for expertise in a domain to be critical was stated as, "If you don't possess the critical competency, you will feel 'daily pain' in carrying out the responsibilities of the billet."[1]

We label the most critical area of domain expertise, the one without which success in a billet is unlikely, as the "primary" domain expertise requirement for the billet. Expertise in the primary domain is so central to performance in the billet that the incumbent must possess sufficient depth of knowledge, skill, and ability as to be identified as having mastered it—the incumbent must have sufficient depth to be an

---

[1] In our original Air Force senior leader development work, we and our Air Force partners struggled to find a definition that would accurately reflect what we meant by "critical." One officer suggested the "daily pain" characterization that we found to resonate with other military officers and that we carried forward into this work.

Table 3.1
Areas of Domain Expertise

	Area of Expertise
Warfighting	Air Warfare
	Expeditionary Warfare
	Information Warfare
	Joint and Combined Warfare
	Mine and Undersea Warfare
	Space Warfare
	Special Warfare
	Submarine Warfare
	Surface Warfare
	Unmanned Warfare
Non-warfighting	Acquisition Management
	Acquisition Professional (Certification)
	Aerospace Engineering and Maintenance
	Civil Engineering
	Counterterrorism
	Cryptology
	Financial Management
	Human Resources
	Information Professional
	Installation Management
	Intelligence
	Logistics and Readiness
	Nuclear Propulsion
	Oceanography
	Operations Analysis
	Public Affairs
	Ship Engineering and Repair
	Strategic Plans and Policy
	Supply Management
	Warfare Resources Management

expert provider of the service. This depth of knowledge, skill, and ability is typically only developed through extensive education, training, and career-long experience.

There are three possibilities with regard to the primary domain requirement related to successful performance in a flag officer billet. First, a billet may require one specific primary domain. For example, some billets may require a submarine warfare officer, and no other primary domain is acceptable. Second, a billet may require one of several different primary domains. For example, some billets' expertise requirements may be equally well met by a surface warfare officer, a submarine warfare officer, or an air warfare officer. Finally, some few flag officer billets may have no requirement for a specific primary domain, i.e., having achieved flag rank provides sufficient expertise to perform successfully in the billet.

Next in importance for successful performance in a billet after the primary domain requirement is the "secondary" domain requirement. The secondary domain is also central to successful performance in the billet. However, the successful incumbent does not need to be a deep expert in the secondary domain, but rather must possess sufficient depth of knowledge, skill, and ability to be identified as one who can perform many of the tasks associated with the domain. This depth of knowledge, skill, and ability is gained through one or more tours in the domain and/or substantial education or training in the domain. Similar to the primary domain, successful performance in the billet requires that the incumbent must have sufficient depth to be a provider of the service,[2] though not with the same level of expertise as if it were the primary domain.

Similar to the primary domain requirement, a billet may require one specific secondary domain, one of several secondary domains, or no secondary domain at all.

Finally, the breadth of responsibility for most flag officer billets requires the incumbent to be familiar with many different domains. We

---

[2] With the exception of some warfighting areas of expertise (air warfare, surface warfare, submarine warfare), the items in Table 3.1 can serve as either primary or secondary areas of domain expertise.

label the required depth of experience in these domains as "familiarity." The incumbent does not need to be an expert in these domains but must possess sufficient depth of knowledge of the domain to be identified as an informed consumer. This level of knowledge, skill, and ability can be attained through education, training, and/or experience.[3]

Leadership, Management, and Enterprise Expertise

Leadership, management, and enterprise expertise are often the first that come to mind when identifying the kinds of expertise that senior leaders need. This is perhaps because flag officer billets are nearly ubiquitous in requiring leadership, management, and enterprise expertise for success (see Chapter Four). However, as noted earlier, possession of expertise in these areas seldom, if ever, is alone sufficient to qualify an individual to be successful in a billet. These kinds of expertise in our framework were organized into the following categories:[4]

- Leading Change
- Leading People
- Stewarding Resources

---

3 Early survey work for the Air Force did not distinguish between primary, secondary, and familiarity areas of domain expertise. We soon realized that general and flag officers, in order to perform effectively in their billets, need to be knowledgeable about a great deal of domain-specific knowledge, but that in most instances that knowledge requires the depth of an informed consumer and not the depth of a deeply experienced provider. Further, it was the judgment of our clients and project staff that officers would not be able to develop service-provider depth in more than two domain areas, and this dictated the single-primary/single-secondary/multiple-familiarity framework that the research employs. Subsequently, however, in reviewing the experiences of RDML selectees, we found several selectees who in our judgment have depth in more than two domain areas.

4 These clusters are similar to those in the leadership model developed by the U.S. Office of Personnel Management (OPM, undated). The OPM leadership model has five clusters: leading change, leading people, building coalitions/communications, results-driven, and business acumen.

- Externally Networking
- Integrating Results
- Accomplishing Mission.

Appendix C contains the detailed list of the specific areas of leadership, management, and enterprise expertise included in our framework. These areas were developed in conjunction with the Navy and were intended to reflect items currently found on flag officers' fitness reports.

Identifying the Requirements for Expertise

In planning our analysis of the expertise requirements of flag billets, we were certain of three things. First, we recognized that the kinds of expertise required for successful performance in flag billets are unlikely to remain completely static over time. Second, flag billets disappear and new billets are added regularly to the roster. Third, we recognized that a single data-gathering cycle was unlikely to be completely accurate in identifying the areas of expertise required across all Navy flag billets. As a result, we undertook our initial analysis with an understanding that the process of determining billet requirements would need to be repeated, either on a regular schedule, such as bi-annually, or on a rotating billet-by-billet basis in which each billet would be reanalyzed six to eight months after a new individual was assigned to it. This meant that, over time, the areas of expertise required for the billets would need to be refreshed.[5]

We employed survey and interview methods for gathering information about the detailed kinds of expertise required for Navy flag billets. The data for this report are the result of the first round of surveys and interviews of flag officers. Survey respondents were flag officers serving in the billets, and most billets had two responses: one from the

---

5 The process of keeping up with billet requirements as they evolve over time may lead to the need for changes in the domain-specific areas of expertise. For example, there were some rather specific domains in the original survey, such as Sea Basing, that were dropped as too specific during the review process.

current incumbent in the billet and one from the immediately preceding incumbent in the billet. Respondents were not asked what their own knowledge, skill, and abilities were; nor were they asked about their own performance in the billet. Rather, the survey questions focused on asking them to identify the critical requirements for successful performance in the billet. A copy of the survey is attached as Appendix D.

The Vice Admiral (VADM) in the billet's chain of command reviewed the responses associated with that billet. For those billets with two responses, the Vice Admiral resolved any discrepancies. In addition, the Chief of Naval Personnel and the Flag Detailer reviewed all the responses.

The Job Book Documents Area-of-Expertise and Other Billet Requirements

The synthesized requirements for each billet, i.e., the requirements that emerged from the review process, are contained in a *job book* that is, as of fall 2007, maintained and electronically accessible through the office of the Executive Learning Officer (ELO). In addition to the area-of-expertise requirements (the domain-specific and leadership/management/enterprise areas of expertise), each billet's entry in the job book also indicates whether the billet has special language or education requirements.

The ELO has responsibility for refreshing and maintaining the job book. We recognize that the passage of time and resulting new challenges will cause a billet's requirements to evolve. The ELO plans to ask each new incumbent to a flag billet, about six months into the new incumbent's tenure in the billet, to review and update the billet's requirements. In this manner, we hope that the job book will continue to accurately reflect the billet's changing requirements.

CHAPTER FOUR

A Closer Look at Expertise Requirements

Chapter Three described our framework for characterizing the expertise requirements of each flag billet in terms of one or more primary areas of expertise, one or more secondary areas of expertise, areas of critical familiarity, and leadership, management, and enterprise areas of expertise. In this chapter, we examine some aggregate characteristics of flag billet expertise requirements and discuss problems associated with these requirements in flowing flag officers through billets.

Flexibility in Primary Area of Expertise

The Navy takes advantage of flexibility of requirements in flag officer billets. One current Navy example is a carrier battle group, in which typically either a surface warfare officer or an air warfare officer is acceptable as commander. As of this date, there has also been at least one example of a submarine officer commanding a carrier battle group, and our survey data illuminate this flexibility. Indeed, our data show substantial flexibility in the primary area of expertise for many such billets. Figure 4.1 summarizes this flexibility.

About 30 percent of flag billets call for a single specific primary area of expertise for successful performance. For example, the billet, "Commander, Submarine Force, U.S. Atlantic Fleet/Commander, Submarine Allied Command, Atlantic," is identified as requiring Submarine Warfare as the primary area of expertise, and no other primary area of expertise will do. About 12 percent of flag billets have the flexibility of accepting one of two different primary areas of expertise. The

Figure 4.1
Flexibility in Primary Expertise Requirements

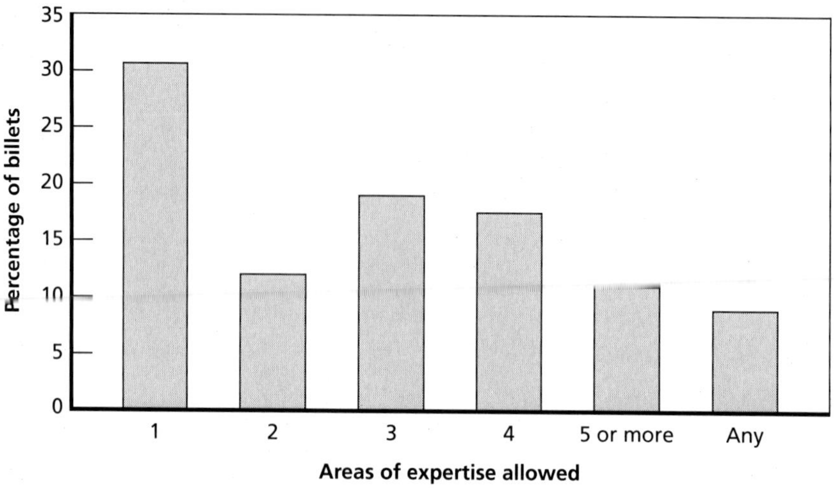

RAND *MG618-4.1*

remaining 56 percent of billets identify three or more primary areas of expertise (including the far right-hand "Any" bar), which provides important assignment flexibility. About 36 percent identify three or four areas of expertise. About 20 percent identify five or more areas of expertise, almost half of which can be filled by a flag officer with any primary area of expertise—"Director, Navy Staff" is one such billet.

Figure 4.2 presents a more detailed perspective of the primary areas of domain expertise, showing the prevalence of these areas of expertise in the data, i.e., the percentage of billets calling for the indicated area of domain expertise—our threshold for including an area of expertise in the figure is that about 5 percent call for that primary area of expertise. The leftmost bar indicates that one in four (about 26 percent) of the billets call for flag officers from any of the Air Warfare, Surface Warfare, or Submarine Warfare communities. The second bar shows that another 19 percent of billets can be occupied by an officer with any one of those three areas of expertise or, additionally, Special Warfare. Note the fifth bar, labeled "Any Primary" and representing nearly 10 percent of the billets: The billets in this group are indifferent to the primary expertise of the officers assigned to them. All together,

Figure 4.2
Most Prevalent Primary Domain Expertise Requirements

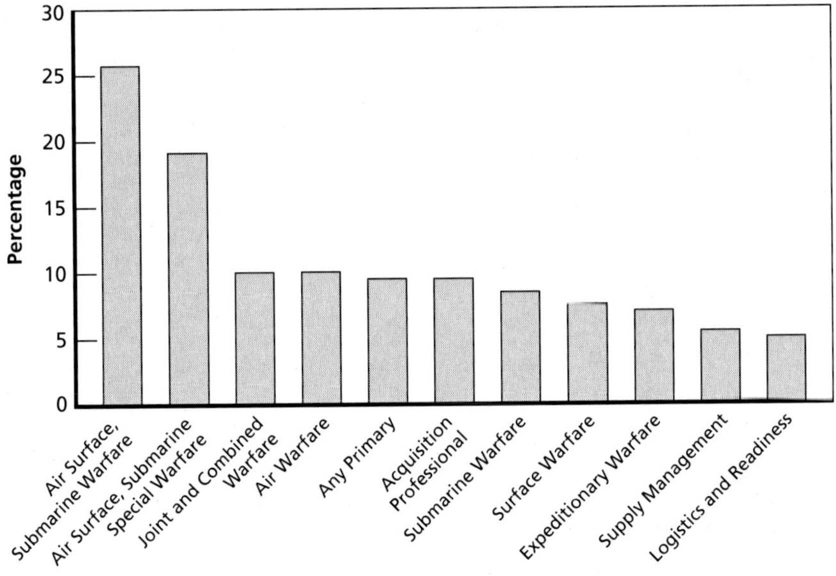

RAND *MG618-4.2*

these three groups of billets make up 65 percent of the positions and provide flexibility in making assignments to mitigate some of the constraints and "bottleneck" assignment issues that we will discuss later.

A substantial percentage of flag billets can be acceptably filled by officers with non-warfighting areas of domain expertise, e.g., Acquisition Professional, Supply Management, and Logistics and Readiness. Unlike the three groupings of expertise mentioned above, there could be overlap among the individual areas of domain expertise shown in the figure. For example, the same flag billet may call for either Acquisition Professional or Logistics and Readiness.

Secondary Areas of Expertise

One issue identified by the CNO when this research began concerned the additional areas of expertise that he believed were required for success in flag officer billets. In Chapter Three, we noted that incumbents were asked to identify either the single second-most important area of expertise for their billet or several areas of expertise, any one of which was important to hold as a secondary area of expertise. Figure 4.3 displays a summary of these data. For example, Financial Management[1] was identified as a secondary area of expertise for over 35 percent of flag billets.[2]

The areas of expertise in Figure 4.3 are the most prevalent secondary areas of expertise for flag officer billets; we use a 10-percent threshold, i.e., at least 10 percent of the positions must mention an area of domain expertise to be included in the figure. Without going any further than this, these data suggest that the Navy would be well served by developing these areas of expertise in its officer corps in addition to the primary areas of expertise it already develops. While there is little doubt that Navy officers already obtain expertise in these areas, these data bolster the case for focused officer development efforts in these areas.

Importance of Leadership, Management, and Enterprise Expertise

In addition to operational and functional areas of expertise, we also gathered data regarding the criticality of a number of specific leadership, management, and enterprise areas of expertise. The percentage of billets for which each was marked critical appears in Table 4.1. Perhaps not unexpectedly, the expertise among these that was most often cited as critical was "exercising responsibility, good judgment, authority, and accountability." Further, of the 38 characteristics listed in Table 4.1, 22 of them are identified by at least 70 percent of the billets, and only

---

[1] Interviews conducted with several flag officers in late 2006 suggest that this area of expertise may be better understood as "strategic management," including expertise in decisionmaking related to strategy formulation, evaluation, implementation, and control, with the latter three addressing the relevant financial management expertise.

[2] These numbers do not add to 100 percent because each billet may have several areas of secondary expertise, any one of which could fill the requirement.

Figure 4.3
Most Prevalent Secondary Domain Expertise Requirements

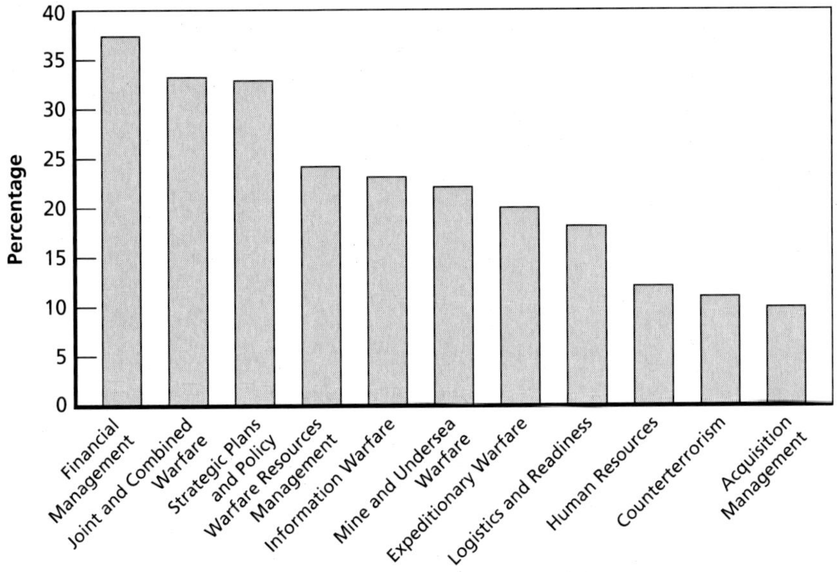

RAND *MG618-4.3*

two are identified by 50 percent or fewer billets. Unlike the primary and secondary areas of expertise, for which flag officers are frequently required to have developed deep knowledge of just two occupational areas of expertise during their pre-flag careers, most flag officers are expected to have developed a wide array of leadership, management, and enterprise expertise during their pre-flag careers.

We wish to underscore what was stated in Chapter One. Unlike the domain-specific areas of expertise, the leadership, management, and enterprise areas of expertise are required of most if not all flag officer positions—the high prevalence percentages in Table 4.1 attest to this. As a result, from an officer-development perspective, the requirements for these areas of expertise are nondiscriminatory and therefore not uniquely relevant for one billet or another. This means, effectively, that all Navy flag officers will need a broad array of leadership, management, and enterprise expertise. However, because the domain-specific areas of expertise do discriminate among flag officer billets, our meth-

Table 4.1
Proportion of Billets Identifying Specific Areas of Leadership, Management, and Enterprise Expertise as Critical

Specific Expertise	Percentage Critical
Leading Change	
Strategic vision to impact the future of the Navy and the nation	82
Strategic vision linking my command or organization to the Marine Corps strategic vision	62
Awareness of world and national affairs that impact the Navy's strategic vision	68
Flexibility to stimulate process development and evaluate new ideas and achieve the Navy's strategic vision	82
Guiding expectations, managing risk, and achieving results	***93***
Engaging in continual learning to master new knowledge	68
Leading People	
Motivating, inspiring, and mentoring military personnel	***95***
Motivating, inspiring, and mentoring civil service employees and contractors	82
Promoting personnel development and team-building	***89***
Resolving conflict and confrontation with and among superiors, peers, and subordinates in a peacetime environment	***91***
Resolving conflict and confrontation with and among superiors, peers, and subordinates in a combat/wartime environment	50
Leading an ethnically and culturally diverse workforce	81
Stewarding Resources	
Knowledge of and effective use of the Planning, Programming, and Budgeting System (PPBS)	***89***
Leveraging technology to achieve business or mission objectives	71
Understanding principles of financial management and marketing	71
Understanding staffing requirements and principles of human resource management	65
Analytical and research abilities to frame problems, synthesize issues, formulate solutions, and select or recommend or courses of action	83
Applying procedures, requirements, regulations, policies, and business principles to make sound resource decisions	78
Externally Networking	
Influencing and negotiating with people at all levels	***91***
Skills necessary to partner with foreign militaries and governments	57

Table 4.1—Continued

Specific Expertise	Percentage Critical
Interpersonal skills necessary to build networks and coalitions	***87***
Preparing and delivering quality oral presentations and written communications	***91***
Building and sustaining effective networks through use of information technology	54
Knowledge of how and who to ask for the capabilities of other Services and Agencies as part of joint operations and warfare	64
Integrating Results	
Skill in employing force to achieve joint, combined, or interagency objectives	47
Knowledge of command and control roles, doctrines, missions, and capabilities of joint/combined forces	56
Resilience and flexibility to deal effectively with change and setbacks	***89***
Exercising good judgment, perception, adaptiveness, and common sense to integrate priorities and eliminate irrelevant information	***95***
Measuring readiness and operational effectiveness to achieve and sustain Joint Operational Excellence	64
Advocating the use of naval forces and sea power within and outside the Navy and Marine Corps	62
Understanding the structures, organizations, capabilities, and cultures of other services	65
Integrating practices and rules of each service to overcome cultural and operational differences and achieve joint/combined objective	49
Accomplishing Mission	
Exercising responsibility, good judgment, authority, and accountability	***99***
Developing and maintaining controls which ensure the integrity of the organization	***87***
Proficiency in problem solving and continuous improvement techniques	***87***
Creating environments where decisiveness and risk management will optimize outcomes and force effectiveness	81
Integrating naval forces into the joint/combined team	57
Ability to provide combatant commanders with flexible, agile, and capable naval forces	62

NOTE: Percentages greater than 85 are in bold, italic type.

odology addresses only the primary and secondary areas of domain expertise, focusing on primary/secondary domain expertise pairs.

Relating Expertise Requirements to Pre-Flag Officer Development

As mentioned in Chapter One, a goal of this research is to suggest domain-specific areas of expertise in which the Navy should focus officer development. Our approach is to frame the question in terms of pairs of primary and secondary domain expertise brought into the flag officer ranks by newly promoted RDMLs. More particularly, the question we seek to answer is: *If the Navy could specify desirable sets of primary and secondary areas of domain expertise, on average, how many promotions to O-7 per year for officers with various combinations of such expertise would the Navy want?*

For this analysis, we assume that the areas of domain expertise under consideration are acquired *prior to* promotion to flag rank and discount the possibility that they may be obtained while serving as flag officers. From the analysis results, we would infer that the combinations of domain expertise so indicated would suggest areas where pre-flag officer development is desirable. Although we are considering actual billets, the officers occupying them in our analysis are virtual rather than actual people. We call the set of hypothetical flag officers in our analysis "inventory."[3] Thus, we answer the above question by undertaking to create a hypothetical inventory of newly promoted RDMLs that, when flowed through the flag ranks, would, if possible, meet all of the requirements for primary and secondary areas of expertise.

From our data, we know what primary and secondary combinations of domain expertise would satisfy the requirements for each billet. For some billets there are several satisfactory combinations of expertise, which provide assignment flexibility. On the other hand, some billets

---

[3] We recognize that important development takes place in the flag ranks and that flag officers do acquire important additional domain expertise while serving in flag billets. However, from the pre-flag officer development perspective, this assumption means that we don't have to depend on development at the flag level to ensure that flag billet requirements are met.

can be filled only with very few alternative areas of expertise. There are cases where domain expertise required for a billet at an upper rank is appropriate for only a few or no positions at lower flag grades. If there are few opportunities at lower grades for officers with the expertise required in higher grades, assignment selectivity will suffer, meaning that choices among candidates to assign to the higher-grade billet could be severely limited. If there are no positions at lower grades that call for domain expertise needed for the higher-grade billet, we would still want to ensure that sufficient candidates for the higher-grade billet have the required expertise. As Figure 4.1 illustrates, there are a large proportion of flag billets that have flexibly specified domain requirements. Our methodology, in constructing the virtual inventory, takes advantage of this flexibility.

Another way of looking at the data is from O-7 upward. There are O-7 billets that could be occupied by RDMLs with several alternative areas of expertise. For example, officers with several different primary areas of expertise could successfully occupy the Commander of Naval Network and Space Operations Command billet. These include Space Warfare, Cryptology, Information Warfare, and any of the three traditional warfighting areas (surface, air, or submarine). However, there are few billets above O-7 that can utilize the first three areas of expertise, so assigning inventory with one of the warfighting areas of expertise would create more flexibility in making assignments at higher grades.

We seek to produce an assignment of inventory to billets assuming the following conditions:

- Primary and secondary areas of expertise of the inventory assigned to a billet match one of the billet's alternatives.
- Acquiring expertise after promotion from Captain to RDML does not occur; i.e., the virtual inventory at the time of promotion to RDML has the requisite domain-specific expertise to meet flag billet requirements.
- Skipping grades is not allowed, meaning that inventory to assign to a billet at, for example, VADM must also be used at grades RDML and Rear Admiral Upper Half (RADM).

Of course, the Navy is under no obligation to observe these conditions in assigning officers to flag billets. But seeing what can be done when meeting these three conditions provides a basis for answering questions related to officer development. Deriving mixes of areas of expertise brought to the flag ranks by cohorts of RDMLs that meet these conditions is not straightforward. In the next chapter, we will discuss a scheme that employs mathematical programming to solve this puzzle.

Flag Officer Requirements Have Critical Bottlenecks

While many flag officer billets can be filled by officers from a broad array of primary and secondary kinds of expertise, a substantial number of billets at each grade have more restrictive expertise requirements. Table 4.2 lists all flag officer billets at the grade of O-8 and above whose requirements are restricted to either one or two primary and secondary areas of expertise, i.e., billets whose expertise requirements are narrow.[4]

The first billet in this table—the O-10 Director, Naval Nuclear Propulsion—calls for an officer whose primary area of expertise is Submarine Warfare, and whose secondary area of expertise can be one of either Nuclear Propulsion or Ship Engineering and Repair. The second billet—the O-9 Commander, Naval Air Forces, U.S. Pacific Fleet—calls for an officer whose primary area of expertise is Air Warfare and whose secondary area of expertise can be one of either Financial Management or Human Resources.

Drilling down on the O-10 billet at the top of the table, we examine other flag officer billets that call for the same combinations of required expertise. While there are a number of lower-grade billets that call for this combination of expertise, these billets can also be filled with other combinations of expertise, as illustrated in Table 4.3. For example, only one O-9 billet can be filled by this combination, but

[4] There are also about 20 RDML positions that call for one or two primary and one or two secondary areas of expertise. We have not included them in the table for reasons of space.

Table 4.2
Flag Officer Billets (O-8 and Above) Restricted to One or Two Primary and Secondary Areas of Expertise

Grade	Billet Title	Primary Expertise	Secondary Expertise
O-10	Director, Naval Nuclear Propulsion, NAVSEA-08	Submarine Warfare	Nuclear Propulsion *or* Ship Engineering and Repair
O-9	Commander, Naval Air Forces, U.S. Pacific Fleet	Air Warfare	Financial Management *or* Human Resources
O-8	Director, Navy and Marine Corps Intranet (NMCI)	Acquisition Professional	Financial Management *or* Information Professional
O-8	Commander, Naval Air Force, U.S. Atlantic Fleet	Air Warfare	Financial Management *or* Human Resources
O-8	Commander, Naval Strike and Air Warfare Center	Air Warfare	Human Resources *or* Joint and Combined Warfare
O-8	Commander, Strike Force Training Pacific	Air Warfare	Surface Warfare *or* Joint and Combined Warfare
O-8	PEO for Strike Weapons and Unmanned Aviation	Air Warfare *or* Acquisition Professional	Financial Management
O-8	PEO for Tactical Aircraft Programs	Air Warfare *or* Acquisition Professional	Financial Management
O-8	Commander, Military Sealift Command	Surface Warfare	Logistics and Readiness
O-8	PEO for Ships	Surface Warfare *or* Acquisition Professional	Financial Management *or* Ship Engineering and Repair
O-8	Commander, Naval Facilities Engineering Command/ Chief of Civil Engineers	Civil Engineering	Financial Management

Table 4.2—Continued

Grade	Billet Title	Primary Expertise	Secondary Expertise
O-8	Commander, Naval Special Warfare Command	Special Warfare	Joint and Combined Warfare *or* Warfare Resources Management
O-8	Commander, Navy Exchange Service Command	Supply Management	Financial Management

there are other acceptable combinations of expertise for this O-9 billet. An O-9 with Surface Warfare/Acquisition Management can also fill the O-9 billet, as can an officer with Surface Warfare/Financial Management or Surface Warfare/Ship Engineering and Repair. Further, an O-9 with Submarine Warfare/Acquisition Management or Submarine Warfare/Financial Management can also fill this O-9 billet. Thus while there are several different combinations of primary and secondary expertise that can fill the O-9 billet, there is only one combination that can fill both the O-9 and O-10 billets.

Only one O-8 billet calls for Submarine Warfare/Ship Engineering and Repair, and only one O-7 billet calls for this combination of expertise. While there are six O-8 billets calling for Submarine Warfare/Nuclear Propulsion, because the O-9 billet does not permit Nuclear Propulsion as a secondary area of expertise, none of these six O-8 billets can be considered feeders to the O-10 billet. Indeed, unless the O-9 billet's expertise requirements are made more flexible to include Nuclear Propulsion as an acceptable secondary expertise, there is only one billet at each of O-7, O-8, and O-9 that can feed the O-10 billet. From the career progression and assignment selectivity perspectives, this is an unacceptable situation.[5] Further, even if the O-9 billet's requirements are made more flexible to include Nuclear Propulsion as an acceptable secondary, there still remains only one O-7 billet with the requisite combination of expertise, also an unacceptable situation.

[5] Having only one billet at each grade, without sufficient care to ensure that officers with the requisite areas of expertise are placed in other billets in those grades, can lead to a very steep pyramid indeed.

Table 4.3
Billets Calling for Submarine Warfare/Nuclear Propulsion or Submarine Warfare/Ship Engineering and Repair

Grade	Billet Title	Primary Expertise	Secondary Expertise
O-10	Director, Naval Nuclear Propulsion, NAVSEA-08	**Submarine Warfare**	**Nuclear Propulsion** *or* **Ship Engineering and Repair**
O-9	Commander, Naval Sea Systems Command	Surface Warfare *or* **Submarine Warfare**	Acquisition Management *or* Financial Management *or* **Ship Engineering and Repair**
O-8	President, Board of Inspection and Survey	Air Warfare *or* Surface Warfare *or* **Submarine Warfare**	**Ship Engineering and Repair**
O-8	Commander, Navy Region Northeast/ Commander, Submarine Group Two	**Submarine Warfare**	Strategic Plans and Policy *or* Mine and Undersea Warfare *or* Installation Management *or* **Nuclear Propulsion**
O-8	Director, Operations and Plans, U.S. Naval Forces Europe/Commander, Submarine Group Eight/Commander, Allied Submarines, Mediterranean	**Submarine Warfare**	Strategic Plans and Policy *or* Mine and Undersea Warfare *or* Installation Management *or* **Nuclear Propulsion**
O-8	Commander, Submarine Group Seven	**Submarine Warfare**	Strategic Plans and Policy *or* Mine and Undersea Warfare *or* **Nuclear Propulsion**
O-8	Commander, Submarine Group Nine	**Submarine Warfare**	Strategic Plans and Policy *or* Mine and Undersea Warfare *or* **Nuclear Propulsion**

Table 4.3—Continued

Grade	Billet Title	Primary Expertise	Secondary Expertise
O-8	Commander, Submarine Force, U.S. Pacific Fleet	**Submarine Warfare**	Mine and Undersea Warfare *or* Logistics and Readiness *or* **Nuclear Propulsion**
O-8	Fleet Maintenance Officer, U.S. Fleet Forces Command	**Submarine Warfare** *or* Surface Warfare *or* Acquisition Professional *or* Ship Engineering and Repair	Financial Management *or* **Nuclear Propulsion**
O-7	Deputy Commander for Undersea Warfare, SEA-07, NAVSEASYSCOM	**Submarine Warfare** *or* Acquisition Professional	Mine and Undersea Warfare *or* Surface Warfare *or* Financial Management *or* Logistics and Readiness *or* Oceanography *or* **Nuclear Propulsion** *or* **Ship Engineering and Repair**

Finally, even with more flexible O-9 billet requirements, the O-8 and O-7 billets in Table 4.3 enjoy substantial assignment flexibility, i.e., they can also be filled by officers with other combinations of primary and secondary expertise, and there can be no guarantee that assignments made today to fill those billets will take into consideration the need for reasonable career progression and assignment selectivity at both O-9 and O-10.

We've drilled down on only the O-10 billet listed in Table 4.2. Other billets in that table have similar characteristics; i.e., while they have constrained expertise requirements, the billets that feed them enjoy considerable assignment flexibility.

This flexibility is certainly useful when Navy leadership works specific assignments, but from a strategic *force structure* perspective, it can pose problems. For example, it is possible that specific assignment decisions made today can foreclose future assignment options and affect future assignment selectivity. While the requirements can shed light on the bottlenecks that might occur, reconciling the myriad options available when considering *all* flag officer assignments requires a strategic perspective and a systematic analytic approach.

Reconciling the Constraining Requirements

The approach described in the next chapter for matching inventory to billets finds a system of career paths that resolves the issues raised by bottlenecked requirements. It takes advantage of billets that have more flexible requirements to structure career paths that meet all billets' expertise requirements and provide good assignment selectivity where possible.

CHAPTER FIVE

Matching Domain Expertise to Billets in the Navy Flag Officer Force

In the previous chapter we characterized assigning virtual inventory to billets under these three conditions as a puzzle:

- Primary and secondary areas of expertise of the inventory assigned to a billet match one of the billet's alternatives.
- Acquiring expertise after promotion from Captain to RDML does not occur; i.e., the virtual inventory at the time of promotion to RDML has the requisite domain-specific expertise to meet flag billet requirements.
- Skipping grades is not allowed, meaning that inventory to assign to a billets at, for example, VADM, must also be used at grades RDML and RADM.

In this chapter, we outline how these three conditions, as well as some additional considerations, are taken into account in a mathematical model that solves the puzzle of finding an incoming inventory that can meet all requirements for expertise in the flag billets. We then present some results. A solution that satisfies the relationships in the mathematical model would indicate acceptable average rates of promotion and acceptable assignment of virtual inventory to billets (i.e., primary and secondary areas of expertise in the virtual people would match one set of acceptable primary and secondary areas of expertise in the billets). This chapter is couched in descriptive terms. For those readers wishing to understand the model's algebraic basis, Appendix E contains a detailed explanation of the mathematical program's formulation. For readers wishing to skip the description of the mathematical formulation entirely, results of the application of the mathematical model begin in the section of this chapter labeled "Some Results from a Sample Run of the Model" and continue in the next chapter.

Modeling the Average Flows of Inventory in the Flag Officer Ranks

For any particular combination of primary and secondary areas of expertise, we are concerned with the average number of admirals with those areas of expertise at each of the four grades and the average number of promotions per year.[1] We assume that the average time in grade developed from historical data will persist in the future and that it is independent of particular areas of expertise but may be different for different grades. A theorem from probability theory (Little, 1961, pp. 383–387), applied to our situation, asserts that for a combination of areas of expertise and grade,

$$\begin{aligned}&\text{Average Number of Admirals in the Grade}\\&= \text{Average Time in Grade}\\&\times \text{Average Promotions to the Grade per Year.}\end{aligned}$$

Other relationships concern the assignment of inventory to fill billets. The data on areas of expertise for the billets are summarized in a matrix (too large to display here) in which we can think of the rows as representing the billets, and the columns as the individual areas of expertise listed in Table 3.1. The entries in the body of the table are "P" for primary, "S" for secondary, or blank. The information in the table is incorporated in the mathematics to specify which areas of expertise can be assigned to each of the billets. Many of the billets indicate that several primary and several secondary areas of expertise are appropriate. We assume that any such primary expertise can be paired with

---

[1] The averages are long-term, steady-state averages under the assumption that the billets will always be as they are in our data.

any of the secondary areas of expertise, and all such combinations are equally acceptable.

"Supply" equations relate assignments to the employment of inventory. For a combination of primary and secondary areas of expertise at a grade,

Total Assignments to Billets = Inventory.

The matrix of billets by areas of expertise is used to make sure that there are no disallowed assignments, that is, that the number of disallowed assignments is zero.

"Demand" equations ensure that billets are filled. For each billet,

Sum over Expertise of Allowed Assignments to a Billet
= Requirement of the Billet.

The requirements of the billets are usually 1, but we sometimes use fractions for joint positions in which the Navy rotates with other services. Even when a billet's requirement is 1, there can be fractional assignments indicating a mixture of areas of expertise that would be assigned to the billet. The fractions are interpreted as proportions of time the billet would be occupied by an officer with the various areas of expertise in the mixture.

Because there is no entry from outside into the flag ranks above RDML, and because numbers of billets decrease with increasing rank, promotion rates decrease also. For a combination of areas of expertise,

Promotions per Year to a Grade
≤ Promotions per Year to the Next Lower Grade.

The ratio of the two quantities in this inequality can be interpreted as a promotion probability. When the promotions per year are aggregated over all areas of expertise, the aggregate number of annual promotions is fixed because the number of billets and average times-in-grade are fixed so that Little's Theorem (Little, 1961) applies. It follows that the aggregate promotion probability for a grade is also fixed. However, we can exert additional control on solutions by multiplying the right-hand side of the relationship above by numbers less than 1. We call these factors "promotion probability upper bounds," which can be tuned by grade and areas of expertise.

Promotion probabilities are inversely related to a notion that has been called "promotion selectivity." Selectivity is about having more people to choose from than there are positions to fill. If for some combination of areas of expertise, the promotion probability to, for example, O-8, is 1, then every O-7 with that combination would be guaranteed a promotion, and selectivity would be nonexistent; i.e., there would be no choice—all with a given combination of areas of expertise who were promoted to O-7 would be promoted to O-8.

It is possible to obtain solutions that have some rather small values for promotion rate variables. For example, a promotion rate equal to 0.05 would be interpreted as one promotion for the type of inventory once every 20 years, on average. To avoid small values, we can set thresholds on promotion rates. Invoking thresholds means that we would allow promotion rates to be zero, but if not zero they would have to be at least as great as the threshold specified for grade of the promotion.[2]

Mathematical programs are designed to optimize a function of the variables. In our case, we associate "costs" to the variables representing the assignment of inventory to billets. In almost all cases, the costs are 1. For some billets, however, any area of expertise—sometimes primary, sometimes secondary, and occasionally both—would be satisfactory. "Any Primary" and "Any Secondary" categories of expertise are allowed to be assigned to those billets. We slightly penalize assignments for specific areas of expertise when an "any" expertise would be allowed for a billet. Then we would know that, for such billets, the assignment of a named area of expertise would have been done to achieve a feasible solution.

---

[2] To implement thresholds requires the inclusion of variables that can only take on values of zero and one. This turns the mathematical program into a "mixed integer program" rather than a more conventional linear program.

Application of the Data in the Model

The analysis in this chapter uses information on 198 billets, 148 U.S. Navy and 50 Joint. They are distributed over grades O-7 through O-10, as shown in Table 5.1. The 50 joint billets are positions that were held by admirals at the time the original survey was made. Since we do not have good information about all the joint flag billets that the Navy might hold, we assume that these 50 are representative of all such positions. As a consequence, and to maintain a reasonable total number of flag officers, we set the requirements for all billets, Navy and Joint, to 1.0.

The set of areas of expertise is listed in Table 5.2. Most can be either a primary or a secondary area of expertise. The second and third columns are counts of how many times each area of expertise is mentioned across the 198 billets. At the foot of the table are sums and averages. On average, there are 1.44 primary areas of expertise mentioned per billet, but the secondary expertise average of 3.33 per billet is more than twice as many. Because of the assumption that all combinations of primary and secondary areas of expertise indicated for a billet are equally acceptable, this implies that there is a great deal of flexibility in how virtual inventory is assigned to billets.

The "Any Primary" and "Any Secondary" areas of expertise appear for billets for which there are no specific areas of expertise indicated. These are treated as actual areas of expertise, but all the other areas of expertise with "Yes" in the fourth (for primary) or fifth (for secondary) columns are also allowed. As mentioned earlier, these extensions to the

Table 5.1
Distribution of Billets Across Grades

Grade	U.S. Navy	Joint	Total
O-7	74	26	100
O-8	47	11	58
O-9	21	10	31
O-10	6	3	9
Total	148	50	198

Table 5.2
Expertise Used in the Model

Expertise	Named as Primary	Named as Secondary	Can Be Primary	Can Be Secondary	Sample Billet[a]
Any primary	19	0	Yes	No	
Any secondary	0	16	No	Yes	
Air, Surface, Submarine, or Special Warfare	38	0	Yes	No	
Air, Surface, or Submarine Warfare	51	0	Yes	No	Primary
Air or Surface Warfare	3	0	Yes	No	
Air Warfare	20	4	Yes	Yes	
Surface Warfare	15	9	Yes	Yes	
Submarine Warfare	17	4	Yes	Yes	
Special Warfare	5	18	Yes	Yes	
Expeditionary Warfare	14	40	Yes	Yes	
Joint and Combined Warfare	20	66	Yes	Yes	Secondary
Mine and Undersea Warfare	0	44	Yes	Yes	
Space Warfare	4	18	Yes	Yes	Secondary
Information Warfare	8	46	Yes	Yes	Primary
Acquisition Professional	19	0	Yes	No	
Logistics and Readiness	9	36	Yes	Yes	
Supply Management	11	3	Yes	Yes	
Civil Engineering	5	2	Yes	Yes	
Aerospace Engineering and Maintenance	2	2	Yes	Yes	
Ship Engineering and Repair	2	10	Yes	Yes	
Intelligence	5	12	Yes	Yes	
Cryptology	4	1	Yes	Yes	
Information Professional	0	13	Yes	Yes	Secondary
Oceanography	0	6	Yes	Yes	
Warfare Resources Management	0	48	No	Yes	Secondary

Table 5.2—Continued

Expertise	Named as Primary	Named as Secondary	Can Be Primary	Can Be Secondary	Sample Billet[a]
Financial Management	3	74	Yes	Yes	
Strategic Plans Policy	8	65	Yes	Yes	Secondary
Human Resources	0	24	Yes	Yes	
Installation Management	1	18	Yes	Yes	
Nuclear Propulsion	1	11	Yes	Yes	
Counterterrorism	0	22	No	Yes	
Acquisition Management	0	20	Yes	Yes	
Operations Analysis	0	10	No	Yes	
Public Affairs	1	8	Yes	Yes	
Unmanned Warfare	0	10	No	Yes	
Total mentions	285	660			
Average mention per billet	1.44	3.33			

[a] The sample billet is the O-9, Commander, Naval Network Warfare Command, billet.

other areas of expertise are penalized in the mathematical program's objective function to avoid having a solution indicate a specific expertise when not really necessary. As it happens, the flexibility permitted by including "Any Primary" and "Any Secondary" is required to achieve feasible solutions.

The third, fourth, and fifth areas of expertise in the list (Air, Surface, Submarine, or Special Warfare, etc.) are combinations of the four areas of expertise below them. There are many positions for which several distinct warfighting primary areas of expertise are allowed. When more than one warfighting area of expertise is permitted for a billet, the model considers all applicable combinations of expertise as well as singletons. These combinations of areas of expertise were introduced to indicate flexibility that exists and to prevent unwarranted specificity. As with "Any primary" and "Any secondary," assignments from these warfighting expertise aggregates are slightly favored in the model solution. For example, a total of 90 U.S. Navy and Joint billets indicate Air Warfare as a possible primary area of expertise. All but

20 of these billets also allow at least one of the other three warfighting areas of expertise. As a result, when Air Warfare occurs in a solution, we know that there is no substitute. The other three warfare areas of expertise are treated similarly.

Deriving Sets of Areas of Expertise from the Billet Database

The rightmost column in Table 5.2 shows the primary and secondary areas of expertise for one billet in the database: the Commander, Naval Network Warfare Command, an O-9 position. Two primary and five secondary areas of expertise are indicated. We assume that, within the set of primary areas of expertise, only one is required and neither is preferred. The same assumptions are made regarding secondary areas of expertise. With each of the two primary areas of expertise paired with each of the five secondary areas of expertise, there are 10 combinations in all, and any one of the 10 pairs is acceptable for filling this billet. The totality of combinations of expertise of this sort over all the billets is the set of combinations of inventory expertise that the model considers.

Average Times in Grade

For the illustrative case discussed below, we set times-in-grade equal to 3.0 years for grades O-8, O-9, and O-10. For O-7, we set it to 3.333 years, which by Little's Theorem implies that average annual promotions to RDML are 30 per year. The O-7 time-in-grade includes time that captains who have been selected for O-7 serve in O-7 billets.

Upper Bounds on Promotion Probabilities and Thresholds on Promotion Rates

Earlier in the chapter, we mentioned placing upper bounds on promotion probabilities and placing minimum thresholds on nonzero promotion rates. The two kinds of constraints interact because the tighter we make the promotion probability constraints, the lower we have to make the promotion thresholds to retain feasibility. And vice versa.

For the illustrative solution discussed below, we set the thresholds on promotion rates for all inventory areas of expertise and grades O-7 through O-10 equal to 0.30 to avoid indications of infrequent promotions that would not seem realistic. The upper bounds on promotion probabilities were 0.75 for O-8, O.60 for O-9, and 0.35 for O-10.[3] These are close to the minimums possible for feasible solutions given the chosen values of 0.30 for the promotion rate thresholds.

Some Results from a Sample Run of the Model

For many billets, there are a variety of suitable expertise combinations. Moreover, when two or more alternative ways of filling a billet exist, there are no expressed preferences.[4] As a result, there are many alternative feasible and equally acceptable solutions. The parameter settings described above do serve to yield solutions that are parsimonious in that the number of different combinations of primary and secondary areas of expertise that appear within a solution is relatively small.

The model is set up to produce a variety of outputs, but the *Assign* and *Promotions* variables are of most interest. *Assign* variables relate areas of expertise to billets and are useful in examining career paths and identifying the set of lower-rank positions that could lead to a higher rank billet. As an example, Table 5.3 shows how the sample billet in Table 5.2 is filled (O-9, Commander, Naval Network Warfare Command, shown in the shaded cells in Table 5.3). There are two sets of areas of expertise that this solution suggests for filling the position. One has primary expertise in Surface Warfare combined with a secondary in Strategic Plans and Policy. The solution indicates that the billet would be filled this way 90 percent of the time. The other way, accounting for the remaining 10 percent, is the combination of expertise that allows Air or Surface or Submarine Warfare as the primary

---

3 A promotion probability of 0.75 to O-8 means that three out of four captains in each cohort of promotees to O-7 will also be promoted to O-8. Setting 0.75 as the maximum permitted promotion probability for promotion to O-8 ensures that no primary/secondary expertise pair will have guaranteed promotion of all its O-7s to O-8, i.e., there will be choice. The 0.6 and 0.35 maximum promotion probabilities to O-9 and O-10 ensure even more choice.

4 As discussed earlier in this chapter, the aggregated areas of expertise (the first five rows in Table 5.2) do enjoy a slight preference over specific and secondary areas of expertise.

Table 5.3
Filling the Commander, Naval Network Warfare Command, Billet

Primary/Secondary Expertise	Grade	Assign	Title
Air, Surface, or Submarine Warfare/ Joint and Combined Warfare	O-10	0.10	CNO
	O-9	1.00	Vice CNO
	O-9	1.00	Commander, Second Fleet
	O-9	0.10	Commander, Naval Network Warfare Command
	O-9	0.28	Director, Joint Staff
	O-8	1.00	Director, N8F, OPNAV
	O-8	0.29	President, Naval War College
	O-8	1.00	Director, N13 OPNAV
	O-7	1.00	Commander, Carrier Strike Group 10
	O-7	0.67	Commander, Carrier Strike Group 3
	O-7	0.65	Commander, Carrier Strike Group 12
	O-7	0.78	Deputy Director, Global Operations, J3 Joint Staff
	O-7	1.00	Commander, Patrol and Reconnaissance Group
Surface Warfare/ Strategic Plans and Policy	O-9	0.90	Commander, Naval Network Warfare Command
	O-8	0.71	President, Naval War College
	O-8	0.74	Director, Joint Theater Air and Missile Defense Organization, Joint Staff
	O-8	0.05	Director, J3 U.S. Strategic Command
	O-7	1.00	Deputy Director, Operations, Allied Maritime Command Northwood
	O-7	1.00	Director, Political/Military Affairs Strategic Plans and Policy J5
	O-7	0.22	Deputy Director, Operations, National Military Command Center J3 Joint Staff

area of expertise, coupled with Joint and Combined Warfare as the secondary area of expertise.

An interpretation of these results is that the combination of Surface Warfare/Strategic Plans and Policy is dominant in that 90 percent of the time this position would be filled by this combination of areas of expertise. This dominance suggests questions that could readily be explored. For example, what if we directed the model to not allow Surface Warfare/Strategic Plans and Policy when filling this position? How would the solution change? Would a feasible solution even exist? If we lost feasibility by precluding the use of Surface Warfare/Strategic Plans and Policy in filling the position, then we have evidence that there is a specific need for officers with primary expertise in Surface Warfare and secondary expertise in Strategic Plans and Policy.

Looking back at Table 5.2, we see that, according to the data, this billet's requirement can also be filled by VADMs with Information Warfare as the primary area of expertise and either Space Warfare, Information Professional, or Warfare Resources Management as secondary areas of expertise; i.e., from the billet's perspective, there is considerable flexibility in filling this billet. Yet the solution does not employ these areas of expertise when creating the notional inventory for this position, and it is possible that there would not be a feasible solution with the position filled in that way.

Exploring Annual Promotion Rates

Of most interest are the *Promotions* variables, particularly for grade O-7, since these may be used to provide insights for officer development. Table 5.4 shows the average annual promotions to O-7 for various combinations of primary and secondary areas of expertise for one solution to the puzzle.

As suggested in the discussion of thresholds on promotion rates and upper bounds on promotion probabilities, a large number of alternative and equally acceptable solutions are possible. The distribution of promotions in the primary areas of expertise in the average O-7 cohort, indicated in the right-hand column in Table 5.4, seems reasonably stable over different solutions. The distributions of secondary areas of expertise (the bottom row in Table 5.4), however, exhibit more

Table 5.4
O-7 Annual Promotions

	Secondary Area of Expertise							
Primary Area of Expertise	Joint and Combined Warfare	Mine and Undersea Warfare	Information Warfare	Logistics and Readiness	Ship Engineering and Repair	Intelligence	Information Professional	Ocean-ography
Any primary			0.3					0.3
Air, Surface, or Submarine Warfare	1.0	1.7						
Air Warfare	1.9							
Surface Warfare			0.3	0.7		0.3		
Submarine Warfare	0.4	1.9	1.9		0.7			
Special Warfare								
Joint and Combined Warfare								
Information Warfare	0.0					0.3	0.3	
Acquisition Professional							0.4	
Logistics and Readiness	0.7							
Supply Management								
Civil Engineering								
Aerospace Engineering and Maintenance				0.3				
Ship Engineering and Repair								
Intelligence	0.9							
Cryptology								
Financial Management								
Public Affairs								
Secondary total	5.0	3.6	2.5	1.0	0.7	0.6	0.7	0.3

Table 5.4—Continued

Primary Area of Expertise	Secondary Area of Expertise								Primary Total
	War Resources Mgmt.	Financial Mgmt.	Strategic Plans	Human Resources	Installation Mgmt.	Counter-terrorism	Acquisition Mgmt.	Public Affairs	
Any primary			0.5						1.1
Air, Surface, or Submarine Warfare		1.3						0.3	4.3
Air Warfare	0.7	1.9		0.7	0.7		0.3		6.1
Surface Warfare		0.7	0.7						2.6
Submarine Warfare		0.7							5.7
Special Warfare	0.7					0.7			1.4
Joint and Combined Warfare			0.7						0.7
Information Warfare									0.6
Acquisition Professional	0.5	0.9							1.8
Logistics and Readiness		0.3							1.0
Supply Management		0.9							0.9
Civil Engineering		0.5			0.4				0.9
Aerospace Engineering and Maintenance									0.3
Ship Engineering and Repair							0.4		0.4
Intelligence									0.9
Cryptology		0.3							0.3
Financial Management	0.7								0.7
Public Affairs		0.3							0.3
Secondary total	2.5	7.8	1.8	0.7	1.1	0.7	0.7	0.3	30.0

NOTE: The shaded cells indicate combinations of primary and secondary expertise that cannot be zero.

variability across different solutions. As we saw in Table 5.2, there are only 1.44 primary areas of expertise indicated per billet on the average, whereas for secondary areas of expertise the average is 3.33. The implication is that, for any given billet, there is much more room for variety in the secondary areas of expertise that are acceptable for filling a billet.

Regardless of how parameters are manipulated to produce alternative solutions, there are several combinations of primary and secondary areas of expertise for which RDML promotion rates cannot be zero. There are nine such combinations that are indicated by the shaded cells in Table 5.4.[5] These will be of interest in the next chapter, in which we examine the areas of expertise possessed by recent RDML selectees.

Figure 5.1 compares the distribution of primary areas of expertise in the right-hand column of Table 5.4 with the distribution of primary areas of expertise across all billets in our requirements database. The distributions are scaled to the respective totals and are expressed as percentages. Figure 5.2 is a companion display showing how secondary areas of expertise are distributed across average O-7 cohorts as compared with the distribution of secondary areas of expertise within the billet requirements data.

There are a few points of interest in the comparisons in Figure 5.1. For example, 8 percent of the billets do not call for a specific area of primary domain expertise. They are flexible billets in that admirals with any areas of expertise can fill them. However, when the model actually determines the areas of primary expertise needed by the inventory to meet *all* billet requirements, we find that only 4 percent of the virtual inventory can have unspecified areas of primary expertise, i.e., higher-grade billet requirements require that lower-grade billets be filled by admirals with specific areas of primary expertise. Because assigning a *specific* area of expertise to those billets is penalized in the model, we infer that assigning the specific area of expertise is necessary for feasi-

---

5 The cannot-be-zero combinations of primary and secondary expertise were determined by executing the mathematical program once for each of the 812 combinations. In each case, the objective function was set to minimize the promotion rate in question. The nine shaded cells in Table 5.4 correspond to the combinations for which a value of zero for the objective function could not be obtained.

Figure 5.1
Comparison of Primary Areas of Expertise in the Data and O-7 Promotions in the Sample Solution

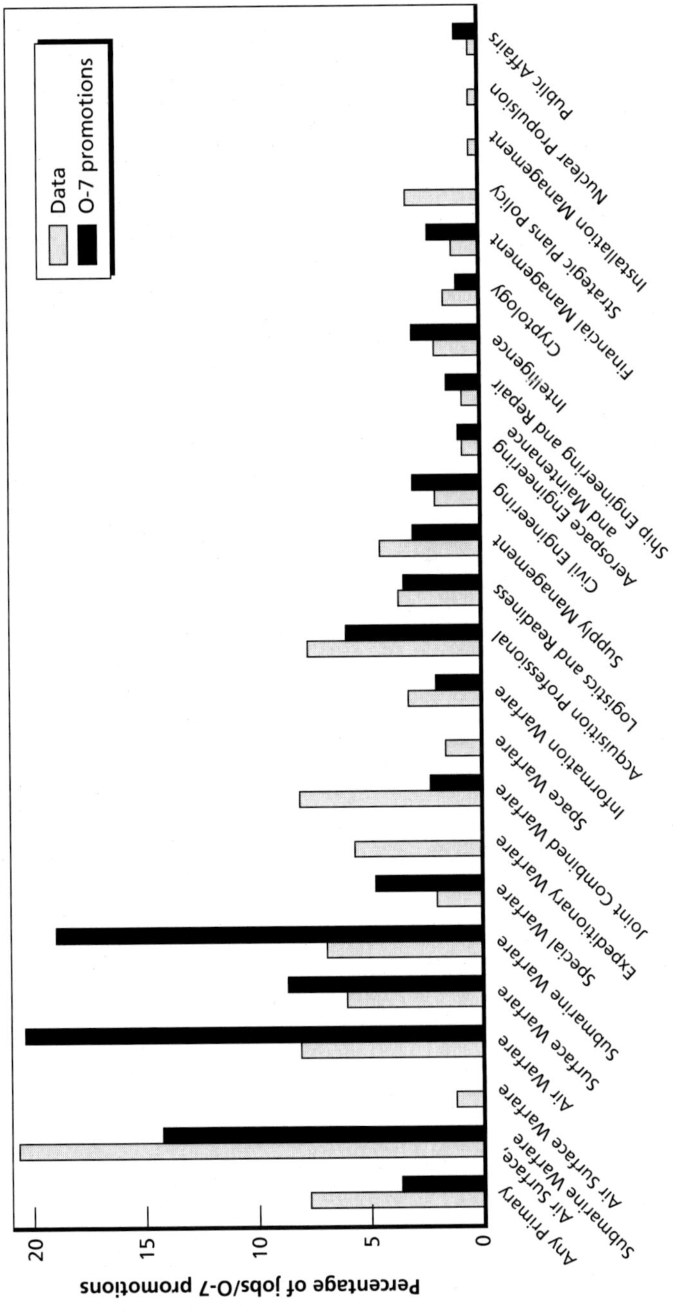

RAND *MG618-5.1*

bility. As another example, we also notice that the Air Warfare, Surface Warfare, Submarine Warfare, and Special Warfare areas of expertise together account for about 64 percent of the average O-7 cohort (rows 2–6 of Table 5.4), but only 45 percent of the total primary areas of expertise indications in the data. This overrepresentation is largely compensated by the sparse appearances of the other warfare areas of expertise in the model solution: Space Warfare, Expeditionary Warfare, Joint and Combined Warfare, and Information Warfare. For the most part, billets indicating these areas of expertise as acceptable also allow one or more of the Air, Surface, Submarine, and Special Warfare areas of expertise. The same is true for Strategic Plans and Policy.

Figure 5.2 mirrors Figure 5.1, displaying the relation between the data and the model solution for secondary areas of expertise. The data show that an individual with a secondary area of expertise of Joint and Combined Warfare could fill slightly over 10 percent of billets. A similar number of billets could be filled by an individual with a secondary area of expertise of Financial Management. From Figure 5.2 (and Table 5.4) we see that these two secondary areas of expertise together account for 43 percent of the average O-7 cohort in the solution. Furthermore, Table 5.4 indicates that seven different secondary areas of expertise cannot be forced to zero in any solution. These are Joint and Combined Warfare, Logistics and Readiness, Ship Engineering and Repair, Financial Management, Strategic Plans and Policy, Counterterrorism, and Public Affairs.

Concluding Remarks

In this chapter, we have employed a mathematical program to help understand some implications of the areas-of-expertise requirements of the billets. The most important finding is that there is a useful amount of flexibility in the expertise requirements. It has not been necessary to make assignment compromises to achieve feasible solutions, i.e., assigning to billets areas of expertise that are not the most appropriate in order to have the areas of expertise available for assignment at higher ranks. Such compromises would be required, however, were we

Figure 5.2
Comparison of Secondary Areas of Expertise in the Data and O-7 Promotions in the Sample Solution

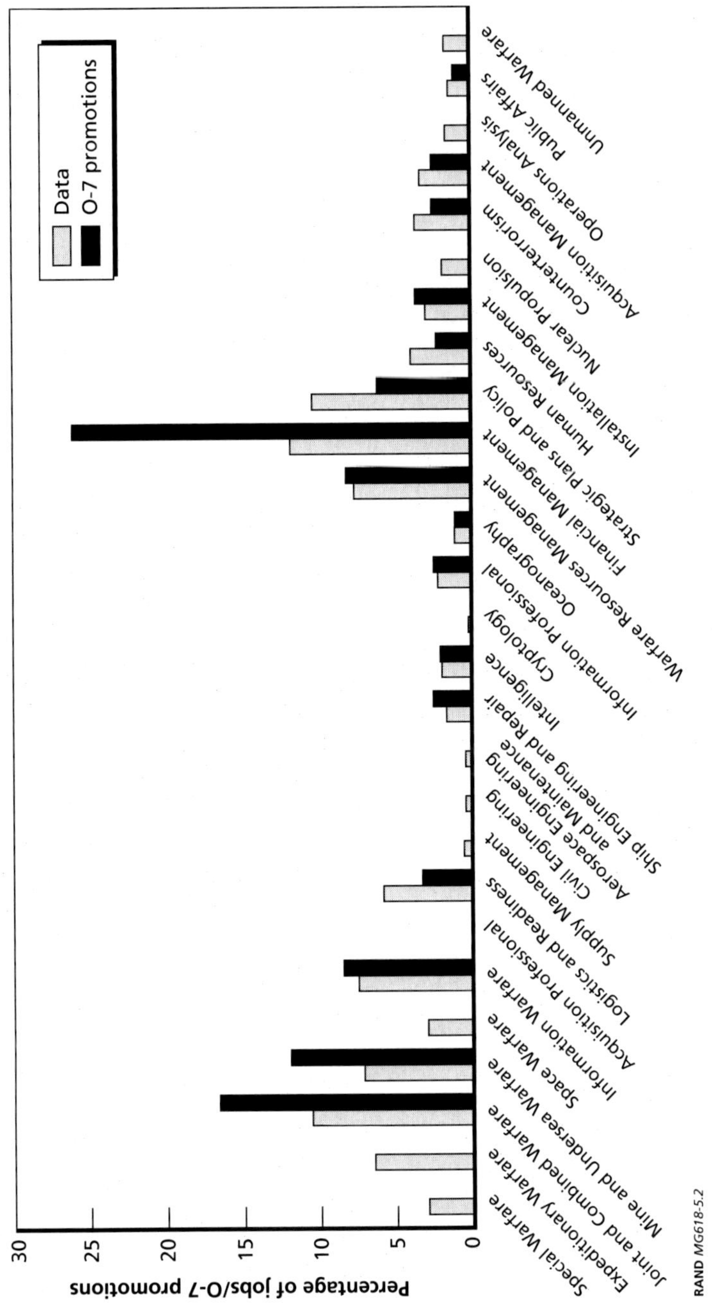

to tighten the promotion probability constraints in order to achieve solutions representing higher degrees of selectivity.

One prescriptive aspect from this exercise is the identification of primary-secondary combinations of areas of expertise that should be present in the flag officer force, i.e., the shaded cells in Table 5.4. We shall return to this notion in the next chapter, in which we make comparisons with recent promotions to RDML.

CHAPTER SIX

RDML Selectees: Comparison with Model-Determined Requirements

The modeling methodology described in the previous chapter provides an ability to assess the areas of expertise represented within annual cohorts of actual RDMLs, as compared to the requirements data discussed in Chapter Two. There are two complementary perspectives. First, are all the pairs of expertise found among actual RDMLs useable in satisfying billet requirements? Second, are all the expertise pairs needed to satisfy billet requirements available in the RDML selections? Seeking answers to these two questions formed the basis of our gap analysis.

Characterizing the Supply

To characterize the supply, we engaged in a multi-step process involving information about RDML selectees from 2001 through 2006. First, using biographies from the Navy personnel Web site, we assigned preliminary primary and secondary areas of expertise, as well as occupational familiarities, to all active RDMLs, RADMs and VADMs. Second, we compared our preliminary expertise assignments with the officers' designators, making adjustments in primary and secondary areas of expertise as needed. Third, using officer record briefs that detail the career points of all flag officers (we were given access to these hard-copy record briefs by the FMO), we performed a consistency check on our expertise assignments, again making adjustments when necessary.

Finally, we restricted our attention only to those flag officers selected on the 2001–2006 RDML selection lists (FMO also provided these lists). In assigning the primary and secondary areas of expertise of the current RDMLs, RADMs, and VADMs from among the 2001 through 2006 selectees, we took care to exclude any experience gained in flag officer assignments; i.e., our assessment only considered experience gained as a Captain or below. This provided the basis for the supply side of our gap analysis. While most of the officers we considered had one primary and one secondary area of domain expertise, we did encounter some officers who in our judgment possessed more than one secondary area of domain expertise. These multiple secondaries were considered in our analysis. Table 6.1 shows 49 primary/secondary pairs of areas of expertise identified among the six cohorts of RDMLs.

Can All the Identified Expertise Pairs Be Used?

Table 6.2 provides a graphic answer to the question of whether there is demand for the supply of expertise pairs. It displays the same information as Table 6.1 in the form of a matrix. Among the 49 cells with entries, reflecting the 49 primary/secondary areas of domain expertise in Table 6.1, 33 have the entry "S/D," and the remaining have "S"—"S/D" implies that there is demand for the expertise pair that matches the supply, and "S" means that there is supply but no demand. For the 49 pairs, we modified the model so that the solution would include as many of the 49 pairs as possible; i.e., we attempted to force the model to utilize the 49 primary/secondary pairs when constructing the virtual inventory.[1] The S/D entries indicate expertise pairs for which we could find solutions that included the indicated pairs. The cells with "S" are cases for which we either could not achieve O-7 promo-

---

1 The efforts included reducing match scores, switching primary and secondary expertise for billet requirements (in this context, switching means that a primary-secondary/secondary-primary match is just as acceptable as a primary-primary/secondary-secondary match), maximizing the number of pairs included, and relaxing promotion probability constraints. Not all these measures were taken simultaneously, and "S/D" indications result from the union of several model runs.

Table 6.1
Area-of-Expertise Pairs Seen in Five Years of Promotions to RDML

Primary Expertise	Secondary Expertise
Air Warfare	Surface Warfare, Expeditionary Warfare, Joint and Combined Warfare, Financial Management, Strategic Plans and Policy, Human Resources, Acquisition Management
Surface Warfare	Submarine Warfare, Joint and Combined Warfare, Space Warfare, Logistics and Readiness, Ship Engineering and Repair Cryptology, Information Professional, Financial Management, Strategic Plans and Policy, Human Resources, Nuclear Propulsion
Submarine Warfare	Joint and Combined Warfare, Logistics and Readiness, Financial Management, Strategic Plans and Policy, Nuclear Propulsion, Acquisition Management
Information Warfare	Space Warfare, Cryptology
Acquisition Professional	Surface Warfare, Submarine Warfare, Space Warfare, Logistics and Readiness, Ship Engineering and Repair, Cryptology, Financial Management, Nuclear Propulsion
Civil Engineering	Acquisition Management
Supply Management	Surface Warfare, Submarine Warfare, Financial Management, Acquisition Management
Aerospace Engineering and Maintenance	Air Warfare, Logistics and Readiness
Ship Engineering and Repair	Surface Warfare, Submarine Warfare, Warfare Resources Management, Nuclear Propulsion
Intelligence	Surface Warfare
Information Professional	Special Warfare, Financial Management, Acquisition Management

tions or the promotion rates were very small. While these officers were selected for flag rank, their primary/secondary areas of domain expertise do not help meet flag billet domain expertise requirements. The half-dozen S indicators in the first three columns reflect an interesting point. In the billet requirements data, we assume that the Air, Surface, and Submarine Warfare areas of expertise can only be primary areas of expertise. However, when reviewing the RDML selection cohorts,

Table 6.2
Primary/Secondary Pairs of Expertise with Supply/Demand and Supply-Only Indications

	Secondary Area of Expertise								
Primary Area of Expertise	Air Warfare	Surface Warfare	Submarine Warfare	Special Warfare	Expeditionary Warfare	Joint and Combined Warfare	Space Warfare	Logistics and Readiness	Ship Engineering and Repair
Air Warfare		S			S/D	S/D			
Surface Warfare			S			S/D	S/D	S/D	S/D
Submarine Warfare						S/D		S/D	
Information Warfare							S		
Acquisition Professional		S	S				S/D	S/D	S/D
Supply Management		S	S						
Civil Engineering									
Aerospace Engineering and Maintenance	S							S/D	
Ship Engineering and Repair		S	S/D						
Intelligence		S/D							
Information Professional				S					

Table 6.2—Continued

Primary Area of Expertise	Secondary Area of Expertise							
	Crypt-ology	Information Professional	War Resources Mgmt.	Financial Mgmt.	Strategic Plans and Policy	Human Resources	Nuclear Propulsion	Acquisition Mgmt.
Air Warfare				S/D	S/D	S/D		S/D
Surface Warfare	S/D	S/D		S/D	S/D	S/D	S	
Submarine Warfare				S/D	S/D		S/D	S
Information Warfare	S							
Acquisition Professional	S			S/D			S	
Supply Management				S/D				S/D
Civil Engineering								S/D
Aerospace Engineering and Maintenance								
Ship Engineering and Repair			S/D				S/D	
Intelligence								
Information Professional					S/D			S

Note: The horizontal line between the fourth and fifth rows separates warfighting from non-warfighting primary areas of expertise.

we found that some selectees had assignments that, in our judgment, provided them with a secondary area of expertise in either Air, Surface, or Submarine Warfare. These were individuals who, in our judgment, possessed sufficient depth in the warfighting areas of expertise even though their primary domain expertise fell in non-warfighting areas.

The full set of expertise pairs where supply cannot be made to fit billet requirements includes

- Aerospace Engineering/Air Warfare
- Air Warfare/Surface Warfare
- Acquisition Professional/Surface Warfare
- Supply Management/Surface Warfare
- Ship Engineering and Repair/Surface Warfare
- Surface Warfare/Submarine Warfare
- Acquisition Professional/Submarine Warfare
- Supply Management/Submarine Warfare
- Information Professional/Special Warfare
- Information Warfare/Space Warfare
- Information Warfare/Cryptology
- Acquisition Professional/Cryptology
- Surface Warfare/Nuclear Propulsion
- Acquisition Professional/Nuclear Propulsion
- Submarine Warfare/Acquisition Management
- Information Professional/Acquisition Management.

A careful review of Table 6.2 shows that supply and demand match in two instances where the selectee has Surface Warfare or Submarine Warfare as the secondary area of expertise: Intelligence/Surface Warfare, and Ship Engineering and Repair/Submarine Warfare. These result from the liberalizing assumption that a primary-secondary/secondary-primary match is just as acceptable as a primary-primary/secondary-secondary match.

Are All Necessary Expertise Pairs Available Among the Six Cohorts?

In Table 5.4, nine cells that cannot be zero are highlighted. Among them, there are only three combinations of expertise that cannot be zero but that we do not see in the analysis of the six years of RDML selectees. These are shown in Table 6.3.

Table 6.3
Missing Primary/Secondary Expertise Pairs

Primary Expertise	Secondary Expertise
Special Warfare	Counterterrorism
Civil Engineering	Financial Management
Intelligence	Joint and Combined Warfare

These three expertise pairs are not represented in the six RDML selection cohorts, and their absence should serve to raise questions about the need for these expertise pairs at the flag officer level. If Navy leadership believes that there is a real need for flag officers with these expertise pairs, then a natural question is to ask about the supply of such officers at the pre-flag grades. If there are officers with these expertise pairs at the pre-flag level, what is it about their experience/background that has kept them from being selected for RDML? More worrisome, if there are no pre-flag officers with these expertise pairs, and if Navy leadership believes that these expertise pairs are important at the flag officer level, what must be done to ensure a reasonable supply of officers for consideration when RDML selection boards meet?

A Final Word on the Gap Analysis

We wish to underscore that the presence of excess expertise pairs (supply) and the absence of needed expertise pairs (demand) does not necessarily indicate a problem at the flag officer level or in the flag assignment process. An officer's areas of expertise are only one aspect when considering that officer for a specific flag assignment. Other, non-area-of-expertise issues may prove much more important when making assignment decisions, e.g., leadership and management skills may dominate, as may other specific requirements known at the time of assignment. The gap analysis should be viewed as providing useful insight but should not necessarily be viewed as reflecting a serious flag officer development or management problem.

Further, the fact that our analysis uncovered so few primary/secondary domain expertise areas where there is demand but not supply demonstrates that the flag force for the most part has the necessary domain expertise to serve effectively in flag billets. For the three domain expertise pairs that are not reflected in the supply, it would be useful to look at the segment of the O-6 population that is viewed as flag-competitive. The presence of these pairs in the flag-competitive O-6 population presents different challenges than the absence of these pairs from that population. If they are present, the question about why they were not selected becomes relevant. Their absence from the flag-competitive population suggests that pre-flag officer development should focus on developing competitive pre-flag officers with these expertise characteristics.

Finally, it is important to underscore that this gap analysis is a qualitative rather than a quantitative one. We did not attempt to judge whether there is sufficient supply to meet demand, but rather whether required domain expertise pairs were present in the 2001–2006 O-7 selectees. This qualitative approach was deliberate. There are many equally acceptable steady-state virtual inventories that can satisfy the flag billet domain expertise requirements. This indeed is a good indication of the robustness of those requirements from the assignment flexibility perspective. We therefore are reluctant to focus on a specific model solution and apply quantitative standards that are unnecessarily restrictive.

CHAPTER SEVEN

An Exploration of Future Requirements

This chapter forecasts the relevance of the areas of expertise in the current framework to the expertise requirements of future U.S. Navy flag officers. The Navy needs to correctly develop its junior personnel of today in the most important areas of expertise needed by Navy senior personnel of the future. The future Navy may acquire new capabilities and objectives that require areas of expertise not recognized by current models of officer development. This forecast is based on an examination of the Navy's structure, its force development, its doctrinal documents, and its technology acquisitions over the past decade and into the next decade. This chapter makes recommendations for some new areas of expertise, some of which are created by further disaggregating or clarifying existing areas of expertise. It addresses two main questions:

- Which areas of expertise are likely to increase in relevance to future senior Navy personnel? This question is important because the Navy needs to correctly develop its junior personnel of today in the most important areas of expertise needed by future Navy senior personnel.
- Are there any areas of expertise that will be important to future Navy billets but are imperfectly captured in the current list? The RAND survey asked respondents to describe the areas of expertise essential to their jobs at the time of the survey. The future Navy is likely to display new capabilities and objectives. These new capabilities and objectives may require areas of expertise not captured perfectly by the existing job book.

How We Derived Future Expertise Requirements

There are no entirely objective indicators by which one could infer future expertise requirements. However, the areas of domain expertise the Navy develops in its officer corps should reflect observable trends in the Navy's future, such as the Navy's future capabilities and objectives. We chose four categories of indicator (or *dimensions*) for these capabilities and objectives:

- *Organizational or structural:* Is the Navy creating new offices or new force structures that emphasize certain areas of expertise?
- *Development:* What areas of expertise is the Navy already emphasizing in its selection, training, or management of junior personnel?
- *Operational strategy:* Is the Navy planning major doctrinal changes or new capabilities or objectives that emphasize certain areas of expertise?
- *Acquisitions:* Is the Navy acquiring or procuring new technologies that emphasize particular areas of expertise?

The four dimensions are intended to cross-check each other. In other words, if evidence for an increased emphasis on a particular area of expertise were found in only one of the four dimensions, then we might conclude that the increased emphasis is slight. In contrast, if evidence were found across all four dimensions, then the increased emphasis should be more dramatic. For instance, the acquisition of some new technology that provides some new capability is likely to have been foretold by an operational strategy document announcing the need for that new capability. However, if an operational strategy document suggests a future capability, yet one observes that none of the necessary technologies are being acquired, then we might conclude that the planning was rhetorical or was tempered by finite resources.

The forecast horizon chosen was the year 2015, 10 years forward from the time of this study. This forecast horizon gives the Navy time to adjust its development of senior personnel, but does not attempt to forecast too far into the future. Long-term forecasts are notoriously unreliable, since plans may be unrealistic or may run afoul of unforeseen future events. Past trends can indicate future trends, although trend analysis is not a perfect science. Good practice suggests that forecasters should attempt to document trends 10 years into the past to validate any trends forecast 10 years ahead.

In order to code the trends for each area of expertise, we examined Navy and DoD documents, mainly organizational and doctrinal documents, as well as journalistic and expert commentary. Since forecasts must be guided by substantive change, we were primarily interested to learn the date on which real organizational, force development, doctrinal, or technological changes were first tangibly observed, rather than when they were first postulated or debated.

Because we present substantial detail in this chapter to support our analysis, we begin first with a summary of our conclusions of what the future holds for changing focus on specific areas of domain expertise. After that we go into specific detail for each dimension about how that dimension reflects changes in Navy focus. For some dimensions, this detail has great depth because of the abundance of concrete examples that reflect emphasis in a particular area of domain expertise. In other instances, there is less depth because of the absence of many concrete examples. The reader wishing to gain an understanding of the trends without delving into the detailed analysis of each dimension need only read this summary.

Summary of Future Significant Impacts on Required Areas of Expertise

The Navy's current job book contains 30 areas of expertise for flag officers. This summary describes which of these areas of expertise are likely to increase in importance to future senior personnel and whether any new areas of expertise should be added to the framework underlying the job book.

We find evidence for the future increased importance of 17 of the existing areas of expertise. Ten show strong evidence in at least two of the four dimensions. These findings strongly suggest that the Navy should pay particular attention to at least the top 10 occupational areas of expertise in the list below during the development of junior staff for senior leadership positions—numbers in parentheses indicate the number of dimensions of strong emphasis:

- Information Warfare (4)
- Surface Warfare (3)
- Submarine Warfare (3)
- Special Warfare (3)
- Expeditionary Warfare (3)
- Intelligence (3)
- Logistics and Readiness (3)
- Air Warfare (2)
- Mine and Undersea Warfare (2)
- Counterterrorism (2)
- Space Warfare (1)
- Unmanned Warfare (1)
- Civil Engineering (1)
- Cryptology (1)
- Financial Management (1)
- Human Resources (1)
- Strategic Plans and Policy (1).

We identified 13 potential new areas of expertise. As with the list of existing areas of expertise, numbers in parentheses indicate the number of dimensions that reflect strong emphasis of these new areas of expertise. Three of them (Homeland Security, Stability Operations, and Theater Ballistic Missile Defense) are strongly emphasized in one dimension but also have some emphasis in two other dimensions—this is why the numbers in parentheses are "1+":

- Information Operations (4)
- Information Technology (4)
- Anti-Submarine Warfare (3)
- Littoral Warfare (3)
- Readiness (3)

- Sea Basing (3)
- Counter-Mine Warfare (2)
- Education and Training (2)
- Homeland Security (1+)
- Stability Operations (1+)
- Theater Ballistic Missile Defense (1+)
- Antiterrorism/Force Protection (1)
- Foreign Areas (as a new career field) (1).

Organizational and Structural Changes

The remainder of this chapter presents evidence across the four dimensions. The first dimension is the "organizational/structural" dimension.

Ship-Related Force Structure

We start by giving an overview of the Navy's overall force structure in terms of the total number of ships and the types of strike groups in which those ships will be organized. In 1991, at the end of the Cold War, the Navy included 538 ships. The 1993 "Bottom-Up Review" concluded that the Navy required 346 ships, 15 carrier battle groups (CVBGs), and 14 carrier air wings. The 1997 and 2001 "Quadrennial Defense Reviews" concluded that the Navy required 305 ships, 12 CVBGs, 10 air wings, and 12 amphibious ready groups. In 2002, Admiral Vernon Clark, the CNO from July 2000 until July 2005, oversaw a major operational strategy document known as Sea Power 21 (Clark, 2002). At that time, the Navy included 318 ships. Sea Power 21 called for a 375-ship Navy by 2020.

In the past, DoD has typically revised the Navy's targets after the Navy has submitted its targets for approval. DoD is then obligated to submit its recommendations to the U.S. Congress. Since Sea Power 21, the House Appropriations Committee has expressed doubt about the justifications for the Navy's most expensive platforms, such as the Littoral Combat Ship, the next-generation destroyer (DDG 1000 *Zumwalt* class) developed under the DD(X) program, and the next-generation cruiser, CG(X). Perhaps reflecting these doubts, in September 2004 the Navy revised downward its shipbuilding program, arguing that the increased capabilities of future ships would allow the Navy to do more with less. On February 10, 2005, the then-CNO told the Senate Armed Services Committee that the increased capabilities of new ships and the exchange of crews at sea ("Sea Swap") would allow the future Navy to get by with fewer than 375 ships. He projected a total of between 260 and 325 ships, even though he also noted that the Navy's current procurement budget could not afford more than 250 ships.

A vigorous debate over the size and composition of the future Navy's fleet continues at the time of this writing. The current proposal envisions a larger Navy of 313 ships by 2036, although this total is not assured. Whatever the total number of ships in the future Navy, the Navy foresees a greater number and variety of strike groups. In 2004, the Navy included 19 strike groups, made up of 12 Carrier Strike Groups and seven Middle East Force surface action groups. Sea Power 21 projected 37 strike groups, made up of 12 Carrier Strike Groups, 12 Expeditionary Strike Groups, nine Strike/Missile Defense Surface Action Groups, and four SSGN (nuclear-powered guided-missile submarine) Strike/Special Operations groups. The latest proposal envisions 11 Carrier Strike Groups, although the numbers of the other groups may not change.

Information Warfare

Turning now to specific areas of expertise, the most notable organizational changes seem to be in Information Warfare (IW). As we shall see, organizational changes alone suggest that IO expertise may need to be disaggregated into two new areas of expertise: Information Operations (IO) and Information Technology (IT). IO is often treated as transitive with psychological operations and strategic communications, roughly meaning "operations intended to influence." IO and Information Warfare (IW) may be treated as transitive, but IW usually includes both influence and counter-influence operations. IT, on the other hand, focuses on network-related warfare, as distinct from IO's emphasis on influence operations.

Since we are documenting trends beginning in 1995, we start with the Fleet Information Warfare Center (FIWC), which was established on October 1st, 1995, as the Navy's "Center of Excellence for Information Operations." FIWC's day-to-day activities include providing IO teams to deploying fleet staffs and naval units; developing IO doctrine and tactics; and providing computer network defense and electronic warfare support throughout the Navy. As of March 2005, FIWC shifted the responsibility for IO away from FIWC N6 (Communication Networks) and placed it with FIWC N3 (Information Operations). Previously, IO and IT were both under FIWC N6. FIWC has been renamed since then as the Navy Information Operations Center. This move from N6 to N3 demonstrates the Navy's change in IO perspective, recognizing that, in today's Navy, information operations are an important component of *operations*.

Other organizational changes in the Navy also suggest an increased operational emphasis on IT. In 1999, a new program executive officer (PEO) for IT was added to the office of the Secretary of the Navy. In July 2002, IT and space technology came under a new command—the Naval Network Warfare Command (NETWAR). NETWAR brought together commands that had been involved in information assurance, information warfare, supporting the fleet, running the Navy's global networks, and running naval space activities. In October 2004, the commander of NETWAR took on the additional and new post of Assistant CNO for IT. In 2006, NETWAR's staff is expected to double from 110 to 220 persons, in part because of an expansion in network defense missions.

Space Warfare

Space Warfare, another named area of expertise in the current framework, also tracks well with the information-related areas of expertise but remains a support, as opposed to an operational, activity. In fact, the Navy's space-related activities are probably not best described as space warfare, since the Navy does not plan to operate any space-based weapons. Rather, the Navy manages a handful of information satellites. The CNO's 2005 guidance refers only to the "joint tactical use of space." The future importance of the Navy's activities in space seems to be a function of the increased importance of the Navy's information-related activities. NETWAR was stood up in 2002 to bring together the Navy's information-related and space-related activities. NETWAR's subordinate Naval Network and Space Operations Command (NNSOC), which was stood up concurrently as NETWAR under an RDML, was formed by the merger of elements of the Naval Space Command and the Naval Network Operations Command. NNSOC operates the Naval Space Operations Center and the Fleet Surveillance Support Command. NNSOC also operates and maintains the Navy's space and global telecommunications systems and services, and directly supports warfighting operations and the command and control of naval forces. Further, a new PEO for C4I and Space[1] was created shortly after the standup of NETWAR.

Intel and Foreign Area Officers

The Navy's space activities support the Navy's intelligence activities, particularly "intelligence, surveillance, and reconnaissance" and "targeting," two of four areas of specialization for the Navy's intelligence community. The Navy has also acquired more "human intelligence" officers in recent years, human intelligence being another of the four areas of specialization. (The fourth area is "intelligence support to special warfare.") Additionally, in April 2005, DoD asked the services to strengthen their programs for foreign area officers (FAOs). The Navy has since decided to establish a career track for FAOs. Until now, the Navy's typical FAO has been an intelligence officer with advanced regional language and cultural skills. FAOs have become more valuable to the Navy due to the increased tempo of coalition operations and the increased DoD-wide emphasis on foreign area skills. One serving Navy officer has written that the Navy should establish a new FAO career track, in part because, at present, FAOs might not exercise their foreign area skills once they return from their short familiarization tour to the

[1] The mission of the Program Executive Officer for Command, Control, Communications, Computers, and Intelligence (C4I) and Space is to acquire, integrate, deliver and support interoperable systems and equipment that enable seamless operations for Fleet, Joint, and Coalition warfighters.

area in question. Two other officers have argued independently that the Navy needs to focus on "operational intelligence," which focuses on the adversary's intentions and capabilities and is therefore related to foreign area expertise.

Terrorism and Counterterrorism

The increased emphasis on intelligence is partly driven by the increased emphasis on terrorism. Antiterrorism or force protection is perhaps captured by the Counterterrorism area of expertise, except that force protection usually refers to defensive measures against terrorism, while counterterrorism is normally understood to also include offensive operations. Since the suicide bombing of the USS *Cole* in Aden, Yemen, in 2002, the Navy has focused more on force protection. The Maritime Force Protection Command (MFPC) was established on October 1, 2004, at Naval Amphibious Base Little Creek in Norfolk, Virginia. MFPC is an element of Fleet Forces Command. The command's mission is to train, equip, and deploy forces to protect Navy units and other assets that are outside secure installations or facilities, as well as to develop their doctrine, tactics, and other standard procedures. Until now, the surface force commanders in the U.S. Atlantic and Pacific fleets had responsibility for force protection units, with the result that force protection competed for resources with the large surface combatants and combat systems. Force protection units include existing naval coastal warfare squadrons (including inshore boat units and mobile inshore undersea warfare units—a total of more than 5,000 naval coastal warfare sailors), explosive ordnance disposal units and expeditionary salvage (diver) units (together, a total of 2,000 sailors, and the recently created Navy Mobile Security Force detachments. One serving Navy officer recommends an officer career track specializing in force protection.

Ballistic Missile Defense

Another area of expertise potentially overlooked in the current framework is theater ballistic missile defense (BMD), which is perhaps captured by the Surface Warfare area of expertise. Theater BMD relies primarily on ship-based anti-missile missiles. Sea Power 21 planned on creating nine Strike/Missile Defense Surface Action Groups by 2020. The eventual total is likely to be less than nine. BMD programs face massive budgetary cuts and still must solve immense technological challenges. While the Navy's theater BMD capabilities are likely to achieve less than originally planned, theater BMD may warrant a separate area of expertise for the Navy to develop, at least because it remains such a large financial investment with important potential.

Anti-Submarine Warfare

Recent organizational changes stress the increased importance of anti-submarine warfare (ASW), which is currently captured within the Submarine Warfare area of expertise, even though ASW involves surface and air warfare, among other things. The CNO formed Task Force ASW in February 2003 within the Navy Staff in Washington, D.C., to review and study options available for revamping the Navy's ASW mission. Initially a study effort, the task force developed into a full-time organization with a core of about a dozen personnel, augmented as needed by the Navy staff.

In April 2004, a Rear Admiral took command of the new fleet anti-submarine warfare command in San Diego, California. The command's primary missions are to foster ASW operations through fleet training, assess ASW performance at all levels through fleet exercises, coordinate with the Naval Personnel Development Center and individual commands in the qualification of ASW personnel, and ensure rapid fleet insertion of advanced technologies. When fully staffed, it will have about 100 personnel in San Diego, 40 in Norfolk, Virginia, and 5 in Yokosuka, Japan.

Special Warfare, Expeditionary Warfare, and Littoral Warfare

The four new SSGN Strike Groups, which are supposed to enter service between 2007 and 2012, add significant stand-off strike and amphibious deployment capabilities to the U.S. submarine force. The SSGN submarines are designed to deploy cruise missiles and special operations teams in particular.

The new SSGN Strike Groups and expeditionary strike groups suggest an increased emphasis on two existing areas of expertise:

"special warfare" and "expeditionary warfare." The new SSGN Strike Groups and the Expeditionary Strike Groups are also supposed to provide new "littoral warfare" capabilities. Littoral warfare is a contested concept but is perhaps most simply defined as "operations where the land meets the sea." The concept is important because such operations can involve stand-off munitions, shallow-water operations, anti-submarine warfare (especially against shallow-water submarines), counter-mine warfare, the landing and support of land forces, upriver operations, and all related air- and logistical-support operations. Littoral warfare is politically salient because states such as China and Iran have increased their own littoral warfare capabilities. The complexity and salience of littoral warfare may warrant a new area of expertise for Navy officers. In late 2002, the Navy changed the phrase "mine and undersea warfare" to "littoral and mine warfare" in the titles of the Deputy Assistant Secretary of the Navy for Mine and Undersea Warfare and the PEO for Mine and Undersea Warfare. Since then, the term "expeditionary warfare" seems to have become ascendant with the formation of Navy Expeditionary Combat Command in late 2005, which brings together the Naval Expeditionary Logistics Support Force, Coastal Warfare Command, Maritime Force Protection Command, ordnance disposal, salvage, and dive assets.

Stability Operations

In January 2005, the then-CNO announced that one of the Expeditionary Strike Groups would act as an "experimental global war on terror expeditionary strike group," which is supposed to increase the Navy's capabilities for "stability operations." "Stability operations" refers to military operations to stabilize and restore regions after war. The current framework contains no area of expertise of that name, although it is probably captured within "expeditionary warfare." Nevertheless, the increased emphasis on stability operations may be sufficient in scale and political salience to warrant a new distinct area of expertise within the Navy.

Readiness, Logistics, Sea Basing, and Training

Increased emphasis on special, expeditionary, and littoral warfare, and new emphasis on stability operations, leads to an increased emphasis on readiness, logistics, and sea basing. These three phrases are captured in the current framework by an area of expertise known as Logistics and Readiness, but this area of expertise may need to be disaggregated. While an increased emphasis on readiness implies an increased emphasis on logistics, readiness seems to include more than logistics, including such things as training and human resources. For instance, in 2003 the Deputy CNO (DCNO) for Logistics (N4) became the DCNO for Fleet Readiness and Logistics, gaining responsibility for fleet and unit training and an increased emphasis on sea basing. Sea basing is certainly captured within the existing Logistics and Readiness area of expertise. However, the increased emphasis on sea basing here and at other points in this chapter suggests that sea basing may itself warrant a dedicated area of expertise of the same name, although some observers doubt whether the Navy has yet invested enough in it.

The increased emphasis on readiness is one reason for an increased emphasis on training. In the previous paragraph, we noted that N4 gained responsibility for fleet and unit training in 2003. At the same time, the DCNO for Warfare Requirements and Programs (N6/N7) gained the former Director for Training organization.

Development Changes

Having examined the main organizational and structural changes, the next (second) dimension we examine covers developmental changes: Is the Navy emphasizing particular areas of expertise in its selection, training, or management of Navy officers?

Shifts in Advanced Education

The Naval Postgraduate School (NPS) is at the center of many notable new opportunities for the development of senior personnel. The existing Financial Management area of expertise is supported by new courses in finance and business management, offered by the Center for Executive Education at the NPS in February 1998. The ELO now supports these activities. Another three courses support information operations, information technology, and information warfare. In February 1998, a new course in information technology was offered at the Center for Executive Education. In 2004, the NPS started an online course in information systems. Another of the courses created at the NPS in February 1998 was a course on space systems, which provides training in space-based surveillance and communications systems, not weaponized space warfare.

Anti-submarine warfare also became the subject of a new course at the NPS in February 1998. In 2003, there was a call for theater ASW commanders to receive more training in ASW, following observations of ASW performance during the Fleet Battle Experiment Kilo of April–May 2003. The observational report concluded that ASW staff also should be increased in number. ASW is not a separate area of expertise in the current framework but is captured by the Submarine Warfare area of expertise, as well as by Mine and Undersea Warfare and Unmanned Warfare.

Surface warfare officers have their own school. This school is one of two functional officer schools commanded by the Naval Personnel Development Command (NPDC), the other being the engineering duty officer school. The NPDC was stood up in 2003 as the result of a deliberate initiative to improve Navy development, known as Task Force EXCEL (2001).

Shifts in Training: High-Tech Ships with Smaller Crews

Training is expected to become more important because of the increasing emphasis on smaller, more skilled ship crews. In 2004, the Navy began its "optimal manning experiments," which experiment with the operation of ships with smaller, more skilled crews. The Navy's next-generation ships are all designed to provide increased capabilities with smaller crews. Navy decisionmakers have repeatedly drawn attention to the Navy's need to retain more skilled personnel and to provide more opportunities for their development. On-the-job training, distance learning, or online courses are intended to provide opportunities for more accelerated and richer personnel development. The Navy terms these initiatives the Total Ships Training Capability.

Training is relevant to the Human Resources area of expertise but perhaps should be captured by a new area of expertise such as "education and training." Navy human resources management is becoming more complicated. The Navy is introducing Sea Swap, which is an initiative to rotate crews at sea, rather than change crews after the ship has returned to port. This initiative allows ships to spend longer periods on station and increases readiness, although Sea Swap may cause unresolved maintenance and other problems for readiness. Some of the Navy's new ships are being designed to support smaller crews and to integrate human resources and technology in more effective ways. The Navy's next-generation destroyer—the DDG 1000 *Zumwalt* class—is being designed explicitly with "human systems integration" in mind, and the next-generation cruiser—CG(X)—is supposed to leverage lessons learned from the DD(X) program.

Operational Strategy Changes

Now we turn to the third of the four dimensions of change: operational strategy. Is the Navy planning major doctrinal changes or new capabilities or objectives that emphasize certain areas of expertise?

Most of the Navy's operational strategy is described in Sea Power 21 (Clark, 2002), a major operational strategy document released in 2002. The document remains relevant, in part because it codified and built on the Navy's previous planning, particularly benefiting from the Navy's early recognition of network warfare and information warfare. However, Sea Power 21 has yet to be developed beyond a short concept paper.

Sea Power 21 (Clark, 2002) described four main concepts to capture the Navy's future capabilities: Sea Strike; Sea Shield; Sea Basing; and ForceNet. For completeness, the original document's definitions of these concepts are summarized below, together with the original lists of their component parts:

Sea Strike is the ability to project precise and persistent offensive power from the sea:

- Persistent intelligence, surveillance, and reconnaissance
- Time-sensitive strike
- Electronic warfare/information operations
- Ship-to-objective maneuver[2]
- Covert strike.

Sea Shield extends defensive assurance throughout the world:

- Homeland defense
- Sea/littoral superiority
- Theater air missile defense
- Force entry enabling.

Sea Basing enhances operational independence and support for the joint force:

- Enhanced afloat positioning of joint assets
- Offensive and defensive power projection
- Command and control
- Integrated joint logistics
- Accelerated deployment and employment timelines.

ForceNet is an overarching effort to integrate warriors, sensors, networks, command and control, platforms, and weapons into a fully netted, combat force:

- Expeditionary, multi-tiered, sensor and weapon grids
- Distributed, collaborative command and control
- Dynamic, multi-path and survivable networks
- Adaptive/automated decision aids
- Human-centric integration.

---

2 Ship-to-objective maneuver is a tactic that consists of using forward-deployed sea basing in international waters to project Marine Air-Ground Task Forces directly to achieve critical operational objectives located deep inland.

There is not sufficient space here to fully document everything relating to Sea Power 21. A full analysis highlights several areas of expertise, as described below, including three (littoral warfare, homeland security, and theater BMD) that are not necessarily captured by the areas of expertise listed in the current framework. We begin with the most emphasized areas of expertise in Sea Power 21: the existing primary areas of expertise, Information Warfare, Intelligence, and Cryptology. New information technologies are supposed to deliver quicker and more effective intelligence. The wider distribution of digital information raises concerns about information access and therefore places an increased emphasis on cryptology. Sea Strike and ForceNet rely heavily on more widely distributed information to enable the highly networked warfare that is the unifying vision for Sea Power 21 as a whole. At the time that Sea Power 21 was released, the CNO wrote that "future naval operations will use revolutionary information superiority and dispersed, networked force capabilities to deliver unprecedented offensive power, defensive assurance, and operational independence to Joint Force Commanders" (Clark, 2002).

Sea Strike relies heavily on surface warfare and air warfare to strike at enemy naval forces and against land-based targets. It also emphasizes the future Navy's enhanced capability to land large amphibious forces to topple regimes or stabilize regions. Sea Strike specifically emphasizes expeditionary warfare, special warfare, littoral warfare (the latter not currently captured directly as an area of expertise), and counter-mine warfare (which is captured within the Mine and Undersea Warfare area of expertise), in part because of the threat to littoral warfare from mines.

Stability operations are clearly related to expeditionary warfare and are emphasized by Sea Strike. Stability operations are given increased emphasis by the CNO's 2005 guidance (U.S. Navy, 2005).

Like stability operations, homeland security is not an operational area directly captured by any of the areas of expertise in the current framework, although it may be included in Counterterrorism. Homeland security is strongly emphasized in Sea Shield. The Navy intends to provide increased support to the Coast Guard to support domestic military operations against major terrorist threats, such as airliners piloted by suicidal terrorists, and to use its theater BMD capabilities to defend U.S. territory against ballistic missile threats.

Finally, Sea Basing is one of the four core concepts of Sea Power 21. Sea Basing refers to accelerated logistics and enhanced readiness. Sea Basing is perhaps the most dynamic of the four Sea Power 21 concepts. Sea Basing currently refers to a group of combatant ships and maritime prepositioning ships, which are meant to provide enhanced fire support and logistical support to land operations without the use of a land base and sometimes without even an aircraft carrier. Sea Basing has some obvious and explicit overlap with expeditionary and littoral warfare, but its major innovation is supposed to lie in enhanced logistical capabilities.

The Navy's 2003 Fleet Response Plan (GlobalSecurity.org, undated) re-emphasizes the objectives behind Sea Basing in Sea Power 21. The Fleet Response Plan aims for a force that is more employable for a longer period of time. (*Employable* refers to the readiness of a force to carry out missions; deployed forces are those actually at sea.) At the time of the release of the Fleet Response Plan, the Navy forward-deployed two or three Carrier Strike Groups and could surge two more within 90 days. The Fleet Response Plan intended to upgrade this "three plus two" capability to a "six plus two" capability: The Navy would be able to deliver six Carrier Support Groups within 30 days, with two more in the basic training phase within 90 days. The Carrier Strike Groups were to be ready within about 72 days as opposed to about 83 days. The old 24-month maintenance, training, and deployment cycle was expanded to a 27-month cycle. In the old cycle, personnel were deployed 25 percent of the time. According to the Fleet Response Plan, personnel are now employable about 55 percent of the time.

Technology Acquisitions

Now we move on to the fourth and final dimension of changes: procurement and acquisitions. Is the Navy acquiring or procuring new technologies, particularly weapon platforms, that emphasize particular areas of expertise?

Information Technology Acquisitions

At the top of the list are a series of early and ongoing acquisitions of information technology. These acquisitions are so numerous that only the most important can be listed here. The Navy began acquiring much of the technology early as part of a foresighted effort known as the Web Enabled Navy. The Joint Tactical Information Distribution System (JTIDS), providing near real-time encrypted voice and data communications, was one of the first components, beginning in 1995 and now in its third phase. However, some of the Web Enabled Navy's components have been delayed or have underperformed. For some critics, the Web Enabled Navy itself remains underdefined. Therefore, it is unclear when the Web Enabled Navy will become a reality, although the effort as a whole is not under threat.

The main shipboard component of the Web Enabled Navy is known as IT21—a ship-wide secure local area network with secure ship-to-shore network connectivity. IT21 has been installed aboard all surface ships, installation aboard all submarines began in 2004 and should be completed in 2011. The Navy/Marine Corps Intranet (NMCI), introduced in October 2003, is the main shore component, replacing hundreds of independent networks with one network. NMCI and IT21 have yet to be integrated. NMCI, at least, is underperforming, due in part to the contractor withdrawing services, claiming financial difficulties. The Navy has admitted that it had contracted poorly.

In 2003, all deploying ships were fitted with secure wide-area networks (SWANs), although the process of certifying their security is ongoing. SWAN will be replaced with the Integrated Shipboard Network System in 2012. Global positioning systems were installed on all ships and aircraft by the end of 2004, somewhat later than expected. Meanwhile, the implementation of the Cooperative Engagement Capability (CEC) is also ongoing. CEC is a system integrating sensor data and weapon systems. In 2004, CEC was installed on 4 carriers, 6 Aegis cruisers, 12 Aegis destroyers, 6 amphibious ships, and 2 squadrons of E-2C Hawkeye 2000 early warning aircraft. The Navy plans to complete installation of its Shipboard Protection System on all vessels bound for the Persian Gulf region in fiscal year 2006 (FY06). The system is a suite of integrated detection sensors, alarms, information displays, and command, control, and communications links.

Submarine Warfare Acquisitions

Submarine warfare is supported by some significant acquisitions, although those acquisitions will provide other capabilities as well. First, four Ohio-class (Trident) ballistic missile submarines (SSBNs) are being converted into nuclear-powered guided-missile submarines (SSGNs), entering service between 2007 and 2012. Each SSGN can carry 154 cruise missiles, 66 Special Operations Forces personnel, and a swimmer lock-out shelter.

The previous CNO once said that the future Navy needs at least 55 attack submarines. At the time of this writing, the Navy plans to deploy a total of 48 attack submarines in 2036, including a total of 10 Virginia-class "New Attack Submarines." USS *Virginia* was delivered in fall 2004, another three have been laid down already, and current budgeting allows for another six. The Navy originally expected to accelerate the procurement of more submarines in this class. However, the Pentagon has curtailed the Virginia-class submarine program. Recently, the Navy has re-emphasized its earlier call for new diesel attack submarines that can operate in shallow waters more easily and quietly than could the Virginia-class submarines.

New attack submarines would enhance the Navy's ASW capabilities. In the surface fleet, ASW will be enhanced by the new Littoral Combat Ship (LCS), the next-generation destroyer (DDG 1000 *Zumwalt* class), and the next-generation cruiser, CG(X), particularly in littoral waters, where the Navy foresees new threats from small, shallow-water, "quiet" diesel submarines, particularly Chinese and Iranian submarines.

All three of the new U.S. classes of ship offer other capabilities, which will be discussed later, but the main description of each class is introduced here. The Navy envisions 55 new LCSs within 30 years, although only 21 have been budgeted so far. The first (USS *Freedom*) was laid down in June 2005. It is scheduled to join the fleet in 2007 and enter service in 2010. The LCS is supposed to utilize a highly automated, small core crew (35–45 persons), high speed (45 knots), and stealthy, shallow-draft (12.8–14.8') hulls. Its "seaframe" design will allow it to accept a number of alternate modules without lengthy refits, such as unmanned undersea vehicles, unmanned surface vehicles, and unmanned and manned vertical take-off and landing vehicles. Its main utility will lie in operations against submarines (primarily quiet diesel submarines), mines, and small surface ships (such as swarming small boats). Officers with the rank of commander likely will be charged with leading the first LCS ships. In the long term, the Navy may assign lieutenant commanders.

Six next-generation destroyers are currently budgeted, the first entering service sometime between 2014 and 2017, with the total production run ranging between 8 and 12. Originally, 24 DD(X) destroyers were expected to enter service. The first design will be built in 2007, two years later than originally planned, following cutbacks in the shipbuilding program, which began in 2001. Subsequent acquisitions may be delayed further. The DDG 1000 *Zumwalt* is intended to utilize all-electric propulsion and auxiliary systems. Electric motors are expected to be more efficient and quieter than current naval power plants and will eliminate the need for hydraulics. The electric power would be charged by gas turbines. Much of its superstructure will be made of graphite composite instead of steel or aluminum. It will have an onboard control system for dispersing heat emissions. Its radars will be embedded in the skin of the ship to make detection by enemy radar more difficult. Instead of the 350 officers and enlisted men and women aboard current destroyers, the DDG 1000 *Zumwalt* class will sail with a crew of 150 or fewer, thanks to increased automation. It will have 80 missile launchers and two main deck guns of 155mm caliber, whose loading and firing will be fully automated.

The next-generation cruiser—CG(X)—would utilize most of the innovations introduced by the DD(X) program. The Navy envisions 15 or 16 ships in the CG(X) class, with production starting in 2019, but the CG(X) enjoys even less consensual support than the DD(X) program.

A rationalized helicopter force and a new helicopter series (MH-60R) deploying improved sensors will also enhance ASW. These helicopters and some other major acquisitions, some of which have been in the pipeline for many years, support air warfare. All the Navy's F14 Tomcats and F/A18 Hornets are to be replaced by F/A18E/F Super Hornet strike fighters by 2010. The Navy will acquire a total of 548 Super Hornets: 222 Super Hornets were procured between 2000 and 2004, with three Squadrons of Super Hornets now in service; 210 Super Hornets will be procured between 2005 and 2009. From 2009 onward, 90 EA-18Gs (the electronic warfare version of the Super Hornet) will enter service to replace the aging, carrier-based EA-6B Prowlers. The Marine Corps STOVL (short take-off and vertical landing) variant of the F-35 Joint Strike Fighter (JSF) will enter service in 2012, eventually replacing all AV-8B Harrier II VSTOL (vertical/short take-off and landing) aircraft. The Navy's carrier version of the F35 will enter service in 2013.

Three of the Navy's 12 in-service carriers are supposed to be replaced by three new nuclear-powered carriers (CVN numbers 77 to 79) before 2020, although only one has been laid down so far. USS *George H. W. Bush* (CVN 77)—a Nimitz-class aircraft carrier—is slated to replace USS *Kitty Hawk* (CV 63) in 2008, although cuts to the shipbuilding budget may delay delivery. The other two of the three new carriers are next-generation aircraft carriers in the CVN 21 class: CVN 78 is programmed to start construction in 2007 and is slated to be placed in commission in 2014 to replace USS *Enterprise* (CVN 65), which will be over its 50-year mark. However, construction may be delayed by cuts in the shipbuilding schedule. CVN 79 is programmed to begin construction in 2012 and to be placed in commission in 2018. Originally, CVN 79 was supposed to replace USS *John F. Kennedy* (CV 67) in the latter's 50th year. However, the December 2004 Program Budget Decision announced that CV 67 would be retired in FY06, reducing the total number of in-service carriers to 11 and putting the future construction of CVN 79 in doubt. The latest annual long-range plan (FY06) suggests that the total number of Navy carriers will never return to 12 carriers and could even fall to 10.

The Navy is reducing the type/model/series of its battle group helicopter force from eight to three: the MH-60 Sierra, MH-60 Romeo, and the MH-53E. This reduction offers efficiencies of simplified maintenance, logistics, and training pipelines. The MH-60R will replace the

SH-60B and SH-60F in the roles of surface and anti-submarine warfare roles. The MH-60S will replace the CH-46D and HH-60H for vertical replenishment, surface warfare, special warfare, and combat search and rescue, and will add an organic airborne mine countermeasures (OAMCM) capability. Ten MH-60S and 10 MH-60R squadrons will deploy as part of the carrier air wings. Detachments from these carrier-based squadrons will in turn support cruiser and destroyer units within the strike group. Consequently, the Navy expects that these helicopter squadrons and their commanding officers will become more closely integrated with Carrier Strike Group leadership. Additionally, five expeditionary MH-60 Romeo squadrons and six expeditionary MH-60 Sierra squadrons will meet non–Carrier Strike Group requirements, such as counter-narcotic detachments, independent cruiser/destroyer operations, search and rescue, and Expeditionary Strike Group support. These expeditionary squadrons are detachment-based units similar to today's helicopter combat support or helicopter anti-submarine (light) squadrons.

The Navy is acquiring a limited number of new unmanned aerial vehicles (UAVs). The Broad Area Maritime Surveillance unmanned aerial vehicle (BAMS UAV) program is intended to develop an unmanned aircraft for future maritime intelligence, surveillance, and reconnaissance missions. The Navy planned to buy 14 BAMS UAVs in the coming years (two in FY07, four in FY08, four in FY09, and four in FY11). But in the FY06 to FY11 budget plan, which the Navy submitted in August 2004, the Navy proposed cutting funding for all except the four in FY11. This would delay the initial operational capability of the BAMS UAV from FY10 to FY13.

Surface Warfare Acquisitions

Surface warfare is supported by several new weapon platforms, on which the Navy will rely as it moves from 7 to perhaps 25 non-Carrier Strike Groups by 2020. The Navy plans to acquire more Arleigh Burke–class (DDG-51) Aegis guided-missile destroyers, for a total of 62. The new DDG 1000 *Zumwalt* class destroyers and the subsequent new-generation cruisers—CG(X)—are supposed to provide enhanced surface warfare capabilities, particularly in littoral waters. In total, the

Navy envisions deploying 88 surface combatant ships by 2036, excluding littoral and expeditionary warfare ships, which are also capable of supporting surface warfare.

Theater BMD will be made real when 18 warships, probably all Aegis destroyers, are equipped with a short- and medium-range engagement BMD capability. The original introduction date was 2005, with completion by 2007, but BMD programs are facing a cut of $5 billion and must still prove their technology. Theater BMD is not currently captured directly by an area of expertise in the current framework, except, perhaps, by Surface Warfare.

Logistics and Readiness

The current area of expertise labeled Logistics and Readiness is enhanced by the acquisition of various new capabilities that made the Fleet Response Plan a reality in 2004. The Navy plans to increase its logistical capabilities so that it can lift three Marine Expeditionary Brigade (MEB) equivalents. Currently, the Navy can lift 2.1 MEB equivalents. The new enhanced lift capability relies on new Amphibious Assault Ships, new Amphibious Transport Docks, and new Maritime Prepositioning Ships, which are also the centerpiece of the new Sea Basing concept. The centerpiece of future sea bases will be 12 new Maritime Prepositioning Force (MPF) ships, to be built beginning in 2007. Components may include a flight deck, accommodation for troops, and a joint command-and-control center. In addition, the Navy plans to deploy 30 logistics vessels and 20 support vessels by 2036.

Expeditionary Warfare Acquisitions

Expeditionary warfare is supported by a number of significant new acquisitions. The Navy envisions deploying 31 expeditionary warfare ships by 2036, excluding the prepositioning, logistical, and support vessels mentioned in the previous paragraph. The last of eight ships in the Amphibious Assault Ship (Multipurpose) or LHD 1 (WASP) class of ships is scheduled for delivery in 2007. The first was delivered in 1989. Beginning in 2013, four LHA Replacement (LHA[R]) ships will replace four USS *Tarawa* (LHA 1) class ships, which will begin to reach the end of their extended service lives in 2011, the last being retired in 2023.

A total of nine San Antonio (LPD 17)-class amphibious transport docks (assault ships) will be built from 2005 onward, with the first entering service probably in 2007. Naval officials once talked of buying as many as 12 LPD-17 ships but, as of the December 2004 Program Budget Decision, the ninth such vessel, budgeted in FY07, becomes the last in the class. The nine ships of the LPD 17 class program will be the replacement for three classes of amphibious ships that have reached the end of their service lives—the LPD 4, LSD 36, and LST 1179 classes—and one class that has already been retired—the LKA 113, replacing a total of 41. The LPD 17 is an assault-transport ship designed to transport aircraft, amphibious vehicles, landing craft, and 700–800 Marines.

While the Navy is enhancing its sealift capabilities, the Navy is cutting its airlift capabilities: Under recent proposed cuts, the Navy would lose 63 C130 transport aircraft, for potential savings of $4.9 billion.

The three new amphibious assault classes—LHD-1, LHA(R), and LPD-17—are supposed to enhance the Navy's expeditionary warfare capabilities. The LPD 17 is the first amphibious ship designed to accommodate the Marine Corps' "mobility triad" for high-speed, long-range tactical-lift operations. The Mobility Triad is composed of Advanced Amphibious Assault Vehicles (AAAVs), Landing Craft Air Cushions (LCACs), and the Marine Corps' new tilt-rotor MV-22 Osprey. The MV-22 Osprey VSTOL aircraft is designed to provide enhanced capabilities for delivering troops and supplies from ship to shore. Originally, a total of 408 Ospreys were slated to enter service from 2006 onward: 360 for the U.S. Marine Corps (USMC) and 48 for the Navy. However, the Program Budget Decision of December 2004 foretold a cut in procurement.

New heavy lift helicopters and new hovercraft will also support expeditionary warfare. The USMC is expected to take delivery of 154 CH-53X heavy lift helicopters from 2013 onward, replacing the aging CH-53E. The CH-53X will carry 32 combat-loaded Marines, or a maximum of 48 persons, or an external load of 27,000 pounds. Although the new CH-53s mission is not entirely competitive with the Osprey, the new CH-53s are not considered as capable as the Osprey. CH-53s are unlikely to join operations within 10 days of orders, a standard that must be fulfilled by all the Navy's platforms by 2015, because CH-53s have to be disassembled, transported to a forward base via cargo planes, and then reassembled before they can support operations. By contrast, MV-22 Ospreys would not need to be disassembled.

A Heavy Landing Craft Air Cushion (HLCAC) is slated for testing in 2009. It would enter service around 2011, if accepted for production. The HLCAC would provide almost double the lift (by area) and would allow faster loading and unloading than the LCAC currently in service.

Littoral warfare will be supported by several new platforms. The LCS, DDG *Zumwalt* class, and the CG(X) (described more fully earlier, under the discussion of anti-submarine warfare) have been advocated in part to fulfill littoral missions. Also, the four new SSGNs are designed in part to support littoral missions, such as landing special operations forces and launching cruise missile strikes against land targets. The LCS, DDG *Zumwalt* class, and the CG(X) are all supposed to incorporate technology developed for the Sea Fighter (previously known as the experimental Littoral Surface Craft or X-Craft), a small, fast landing craft launched in April 2005. Although a transformational program, Sea Fighter indicates the design and capabilities of the later littoral platforms. Concept exploration for the Sea Fighter was first contracted in 1998. Sea Fighter is a high-speed aluminum catamaran, with propulsion by water-jets driven by gas turbines for high-speed operation (50 knots or more) and diesel engines for lower-speed loitering. The vessel measures approximately 73 meters in length and 22 meters in breadth and has a design displacement of approximately 1,100 long tons. Although small, it is self-deployable (4,000 nautical miles unrefueled). The flush upper deck has a landing area for two helicopters, which can also be refueled with the craft under way. Access to the main payload deck is via a large lift down from the flight deck or over folding ramps at the stern. Sea Fighter may also carry the "Affordable Weapon System" (featuring cruise missiles that, at a cost of about $100,000 each, are relatively inexpensive), which would give Sea Fighter deep-strike capabilities.

Counter-Mine Warfare Acquisitions

Counter-mine warfare is a major concern of littoral warfare. Many of the new littoral platforms are supposed to provide enhanced counter-mine capabilities. Additionally, the two experimental high-speed vessels (HSVs), which began operational tests in 2003, will provide increased mine warfare capabilities, primarily through the use of new sensors. The HSV has seen operational service in the Gulf and in relief operations following the Southeast Asian Tsunami of December 26, 2004, where its speed, shallow draft, and helicopter platform drew praise.

Many new mine sensors are being developed for installation on the new littoral platforms. Before then, in 2007, the Remote Mine hunting System (RMS), an unmanned underwater vehicle (UUV), operating the AQS-20A mine-hunting sonar, will deploy from surface ships to search for mines. Five mine sensors are to be installed on the Navy's helicopters between 2007 and 2010, according to the most recent programming decisions (March 2005), providing an OAMCM capability:

- The AQS-20A, the advanced mine-hunting sonar also operated by the RMS, will operate from the MH-60S helicopter in 2007.
- The Airborne Laser Mine Detection System (ALMDS) promises a much faster means to find moored sea mines. The ALMDS is scheduled for fielding in 2008.
- The Airborne Mine Neutralization System (AMNS) is a helicopter-towed weapon, which locates and destroys moored mines. The AMNS is scheduled for fielding in 2008.
- The Rapid Airborne Mine Clearance System (RAMICS) is a laser-aimed gun to defeat mines on or near the surface. The RAMICS is scheduled for fielding in 2010.
- Finally, the fifth of the five new helicopter-mounted mine-countermeasure systems is the Organic Airborne and Surface Influence Sweep (OASIS), a towed device that imitates a ship's

magnetic and acoustic signals to set off influence mines. The OASIS is scheduled for fielding in 2008.

Special Warfare Acquisitions

The two experimental HSVs, whose missions include fast delivery of SEAL teams, support special warfare. The SEAL force itself will expand with the acquisition of two new SEAL teams. The operational date of these new teams is unknown.

Stability Operations: No Acquisitions Directly Relate

Some of the new acquisitions listed under logistics, expeditionary warfare, littoral warfare, and special warfare clearly have relevance to stability operations. However, no currently planned acquisitions are specifically designed with stability operations in mind, although the most recent CNO's 2005 guidance (U.S. Navy, 2005) claims that new ship designs will seek to support stability operations.

The Relevance of Current Expertise for the Future

Of the areas of expertise in the current framework, 17 show at least some evidence for increased relevance. Ten show strong evidence in at least two of the four dimensions. These findings strongly suggest that the Navy should pay particular attention to these 10 occupational areas of expertise during the development of junior staff for senior leadership positions—numbers in parentheses indicate the number of dimensions that reflect increased emphasis:

- Air Warfare (2)
- Counterterrorism (2)
- Expeditionary Warfare (3)
- Information Warfare (4)
- Intelligence (3)
- Logistics and Readiness (3)
- Mine and Undersea Warfare (2)
- Special Warfare (3)

- Submarine Warfare (3)
- Surface Warfare (3).

Information Warfare is the only area of expertise that shows increased emphasis in all four dimensions. Previous discussion suggested that this area of expertise may need to be disaggregated into two separate areas of expertise: Information Operations and Information Technology. Expeditionary Warfare, Intelligence, Logistics and Readiness, Special Warfare, Submarine Warfare, and Surface Warfare all show evidence for increased importance in three of the four dimensions we examined. Air Warfare, Counterterrorism, and Mine and Undersea Warfare all show increased emphasis in two dimensions. Seven areas of expertise show some evidence for increased importance, but show evidence for increased importance in the future in only one dimension: Civil engineering, Cryptology, Financial Management, Human Resources, Space Warfare, Strategic Plans and Policy, and Unmanned Warfare.

Potentially Uncaptured Areas of Expertise

We identified 13 potential new areas of expertise to include in the framework: Information Operations, Information Technology, Anti-Submarine Warfare, Littoral Warfare, Readiness, Sea Basing, Counter-Mine Warfare, Education and Training, Homeland Security, Stability Operations, and Theater BMD. There is at least some evidence for increased relevance across all of these 13 areas of expertise:

- Information Operations and Information Technology are two potential new areas of expertise resulting from the desegregation of the secondary area of expertise named "information operations and information technology."[3] These two are the only potential

[3] As stated earlier in this chapter, Information Operations and Information Technology are disaggregated for two reasons: They reside in different Navy organizations (N3 and N6, respectively), and they focus on distinctly different areas. Information Operations aim at influencing, while Information Technology aims at network warfare.

new areas of expertise that show increased emphasis in all four dimensions.

- ASW, Littoral Warfare, Readiness, and Sea Basing show evidence for increased importance in three of the four dimensions.
- Counter-Mine Warfare and Education and Training show strong evidence across two of the four dimensions.
- Homeland Security, Stability Operations, and Theater BMD are three potential new areas of expertise that show strong evidence in one dimension and some evidence in two other dimensions.
- Antiterrorism or force protection and foreign affairs (as a new career field) show strong evidence in one of the four dimensions.

CHAPTER EIGHT

Conclusions and Recommendations

Conclusions

Admiral Clark was right to be concerned for the development of Navy flag officers and the demands placed on them. Not because there is failure afoot, but because opportunities for improvement abound in a time of fast-moving technology and constrained resources. In addition to the requirements for expertise in the community or designator code normally associated with each billet, our data show that flag billets also require depth of expertise in domains not always deliberately developed in individuals who rise to flag rank. The secondary domains of expertise that billets most commonly require are shown in Figure 4.3 and listed here in order of their prevalence in billet requirements:

- Financial Management
- Joint and Combined Warfare
- Strategic Plans and Policy
- Warfare Resources Management
- Information Warfare
- Mine and Undersea Warfare
- Expeditionary Warfare
- Logistics and Readiness
- Human Resources
- Counterterrorism
- Acquisition Management.

Our data suggest that development in these domains should be widespread for those likely to become flag officers.

Additionally, we found that today's flag officer force does not reflect important domain expertise pairs that, from our data, appear necessary for the Navy to acceptably fill all of its flag positions. The results of our modeling, coupled with our analysis of the areas of expertise reflected in six years of RDML selectees, identified the following required pairs of expertise as lacking among flag officers:

- Special Warfare/Counterterrorism
- Civil Engineering/Financial Management
- Intelligence/Joint and Combined Warfare.

We do not intend to suggest that flag officers have failed in their assignments as a result. Indeed, our analysis suggests that, for the most part, the primary/secondary areas of domain expertise possessed by recent RDML selectees match well with flag billet requirements. Further, other RAND research (Scott et al., 2007) finds that successful leaders develop mechanisms for coping with shortfalls in domain expertise. The much more likely result of lacking required domain expertise is inefficiency—a significant proportion of a flag officer's relatively brief time in a billet is spent coming up to speed in, or compensating for, those areas where expertise is lacking. As Gabarro (1987) noted, "prior functional experience does matter, and it influences how a manager takes charge, the areas he is most likely to deal with effectively, and what problems he faces as he takes charge" (p. 68).

As for future requirements, our review of expertise required into the future found strong evidence for the growing importance of expertise in 10 areas: Air Warfare, Counterterrorism, Expeditionary Warfare, Information Warfare, Intelligence, Logistics and Readiness, Mine and Undersea Warfare, Special Warfare, Submarine Warfare, and Surface Warfare. The Navy currently has mechanisms to develop officers with many of these areas of domain expertise, e.g., Submarine Warfare and Surface Warfare, but other areas may require increased development efforts, e.g., Information Warfare and Intelligence.

The details of what it means to have expertise in some of these domains remain to be fleshed out. For example, we know from our survey data that flag officers consider expertise in "financial management" critical to success in their billets. But what does expertise in financial management consist of for flag officers whose roles are not primarily financial management but who lead large Navy business units or organizations? Recent flag officer interviews suggest that "strategic management" is a more appropriate way to characterize this expertise.[1]

In addition to developing an understanding of what expertise in some of these domains means (e.g., financial management, installation management), how can the Navy identify or create expertise in these domains in its flag officer ranks? For example, should a flag officer who completed an MBA as a lieutenant be considered to have strategic management expertise sufficient to perform successfully in a flag billet? Is participation in a brief ELO-sponsored executive business course sufficient to substantially shorten the time it takes to come up to speed in an unfamiliar domain area? These questions were not addressed by this research and remain unanswered.

With regard to other kinds of expertise critical to flag officer performance, our data clearly indicate that flag officers need to have expertise in leadership, management, and a depth of knowledge of and experience in the Navy enterprise. The three most widely cited critical areas of expertise of this kind are:

1. exercising responsibility, good judgment, authority, and accountability
2. motivating, inspiring, and mentoring military personnel
3. exercising good judgment, perception, adaptiveness, and common sense to integrate priorities and eliminate irrelevant information.

---

1 Flag officer interviews in late 2006 suggest that it may be expertise in strategic management that is needed and not expertise in traditional financial management topics such as corporate finance and capital budgeting. Strategic management includes expertise in decision-making related to strategic formulation, evaluation, implementation, and control.

Our research did not undertake to assess the degree to which the recent cohorts of RDML selectees have these qualities, but the military services place great effort toward developing these qualities in their ranks, so we suspect that there is no shortage of the above kinds of expertise among the flag ranks.

Finally, our analysis rests heavily on a solid understanding of the requirements for successful performance in each flag billet. An important aspect of our research was the development and refreshing of the job book, a document that reflects the domain-specific and domain-independent requirements of each flag billet. Considerable effort was devoted to the development of the job book: (1) We initially surveyed at least two incumbents for each flag billet, the current incumbent and the immediate past holder of the billet; (2) survey responses were synthesized, checked for consistency with other similar flag billets, and then reviewed and adjusted by the VADM in the billet's chain of command; (3) the Chief of Naval Personnel then reviewed each billet's requirements and made additional adjustments as he deemed appropriate; and (4) a mechanism was developed for and is being administered by the ELO to periodically update each billet's requirements, with requirement updates taking place at about six months into a new incumbent's tenure in the position. We believe that this systematic review and update makes the job book a valuable resource that has the potential to guide the development and assignment of senior leaders.

Recommendations

Perhaps the most important result of our research is that it demonstrates that flag officers need to have depth of domain expertise beyond that associated with their officer designator code. Our recommendations are focused on the importance of developing pre-flag officers who are competitive for future flag rank to have specific primary/secondary areas of domain expertise, with consideration for how certain areas of domain expertise will increase in importance over the next decade. Doing this requires an up-to-date understanding of the expertise that is needed and of using that information in the development and assignment of flag officers.

In order to best prepare Navy officers to fill flag officer billets, either in terms of career-long development or in terms of just-in-time development, it is important that the Navy maintain an understanding of the domain expertise that is required. The Navy cannot rely on leadership skills alone, nor on a single officer designator code to convey the additional depth and breadth of expertise required to fill its most senior leadership billets. It is therefore very important to maintain an up-to-date database of requirements as those requirements over time. We see this database or job book as an organic document and recommend that it be maintained by the Navy ELO through the course of regular interactions with flag officers as they move from billet to billet. This mechanism would allow those responsible for development to maintain a close connection with the development needs of flag officers.

Our research has only scratched the surface with regard to a more detailed specification of the expertise required. More needs to be understood about the content and depth of development that is required to drive the focus of development opportunities. For example, would the programs currently offered by ELO be more useful if each program had a narrower focus in terms of both students and material? This suggests to us the need to develop and maintain feedback loops among ELO, the flag detailers, and the flag officers that focus on understanding the demands for expertise in flag billets.

Finally, gaining a better understanding of the expertise required can be used to illuminate the kinds of expertise that are better developed earlier in a career and over a longer period of time and the kinds of expertise that are acceptably developed in a just-in-time context. For example, perhaps some kinds of business-related expertise should be more broadly developed among lieutenant commanders and commanders who can put it to good use in staff assignments throughout the Navy's shore establishment. This would have the added advantage of laying the groundwork for the more in-depth development that might be required for more senior billets.

APPENDIX A

Flag Billet Titles

This appendix lists the billet titles for all the flag officers listed on the United States Navy Biographies Web page (U.S. Navy, undated), accessed May 8, 2007.

Admiral

Chief of Naval Operations
Commander, U.S. Central Command
Commander, U.S. Fleet Forces Command
Commander, U.S. Naval Forces Europe; Commander, Allied Joint Force Command Naples
Commander, U.S. Pacific Command
Commander, U.S. Pacific Fleet
Commander, United States Southern Command
Director, Naval Nuclear Propulsion
Vice Chairman of the Joint Chiefs of Staff
Vice Chief of Naval Operations

Vice Admiral

Chief of Naval Personnel; Deputy Chief of Naval Operations (Manpower, Personnel, Training & Education)
Chief of Navy Reserve; Commander, Navy Reserve Force

Commander, Naval Air Forces; Commander, Naval Air Force, U.S. Pacific Fleet
Commander, Naval Air Systems Command
Commander, Naval Forces Central Command; Commander, Fifth Fleet; Commander, Combined Maritime Forces
Commander, Naval Network Warfare Command
Commander, Naval Sea Systems Command
Commander, Naval Surface Forces; Commander, Naval Surface Force, U.S. Pacific Fleet
Commander, Navy Installations Command
Commander, Second Fleet; Director, Combined Joint Operations from the Sea Center of Excellence
Commander, Seventh Fleet
Commander, Sixth Fleet; Deputy Commander, U.S. Naval Forces Europe; Commander, Joint Forces Maritime Component Command, Europe; Commander, Naval Strike and Support Forces NATO; Commander, Allied Joint Command Lisbon
Commander, Submarine Force; Commander, Submarine Force, Atlantic; Commander, Allied Submarine Command; Commander, Task Forces 46, 82, 84, 144
Commander, Third Fleet
Deputy Chief of Naval Operations for Communication Networks (N6)
Deputy Chief of Naval Operations for Fleet Readiness and Logistics (N4)
Deputy Chief of Naval Operations for Information, Plans and Strategy (N3/N5)
Deputy Chief of Naval Operations for Integration of Capabilities and Resources, OPNAV N8
Deputy Commander, U.S. Central Command
Deputy Commander, U.S. Fleet Forces Command
Deputy Commander, United States Transportation Command
Deputy Director, Central Intelligence Agency
Director for C4 Systems (J6), The Joint Staff
Director for Force Structure, Resources, and Assessment, J8, The Joint Staff

Director, Material Readiness and Logistics (N4), OPNAV Staff
Director, National Geospatial Intelligence Agency
Director, Navy Staff
Naval Inspector General
Principal Deputy Director, Program Analysis and Evaluation, Office of the Secretary of Defense
Superintendent of the United States Naval Academy
Surgeon General; Chief, Bureau of Medicine and Surgery
United States Representative, North Atlantic Treaty Organization Military Committee

Rear Admiral

Assistant Chief of Staff, Navy Expeditionary Logistics Response Center
Assistant Chief of Staff, Navy Medicine National Capital Area
Assistant Commander for Logistics and Industrial Operations
Assistant Commander, Navy Personnel Command for Career Management (PERS-4)
Assistant Deputy Chief of Naval Operations for Information, Plans and Strategy (OPNAV N3/N5B)
Assistant Deputy Chief of Naval Operations for Integration and Augmentation (OPNAV N3/N5 I)
Assistant Deputy Chief of Naval Operations for Resources, Requirements and Assessments (N8R)
Assistant Deputy Chief of Naval Operations (Manpower, Personnel, Training and Education)
Assistant Deputy Chief of Staff for Logistics, Fleet Supply and Ordnance, Commander, Pacific Fleet; Commander, Logistics Task Force Pacific; Commander, Naval Logistics Forces Korea
Assistant Deputy Judge Advocate General; Deputy Commander, Naval Legal Service Command
Assistant Deputy Surgeon General for Total Force Integration; Associate Deputy Chief, Bureau of Medicine and Surgery, Human Resources

Associate Director of Naval Intelligence
Associate Director, OPNAV N81D
Battle Staff Director (J3R), United States European Command
Chief Engineer, Space & Naval Warfare Systems Command
Chief of Information
Chief of Legislative Affairs
Chief of Naval Air Training; Commander, Navy Region South
Chief of Naval Research; Assistant Deputy Commandant of the Marine Corps for Science and Technology; Director, Test and Evaluation and Technology Requirements
Chief of Navy Chaplains
Chief of Staff, Navy Enterprise (N09X)
Chief of Staff, Navy Reserve Forces Command
Chief of Staff, Navy Security Group Reserve Command
Chief of Staff, U.S. Joint Forces Command
Chief of Staff, U.S. Pacific Command
Chief, Strategy Division, J-4, The Joint Staff
Commandant, Industrial College of the Armed Forces, National Defense University
Commandant, Naval District Washington
Commander, Abraham Lincoln Strike Group
Commander, Carrier Strike Group 2
Commander, Carrier Strike Group 3
Commander, Carrier Strike Group 5; Battle Force 7th Fleet; Task Force 70/75
Commander, Carrier Strike Group 7; Commander, USS Ronald Reagan Strike Group
Commander, Carrier Strike Group 8; Commander, Dwight D. Eisenhower Carrier Strike Group
Commander, Carrier Strike Group 9
Commander, Carrier Strike Group 10
Commander, Carrier Strike Group 11, Nimitz Carrier Strike Group
Commander, Carrier Strike Group 12, Enterprise Strike Group
Commander, Center for Submarine Counterterrorism Operations
Commander, Combined Joint Task Force, Horn of Africa

Commander, Defense Supply Center Richmond
Commander, Defense Supply Center, Columbus, Defense Logistic Agency
Commander, Expeditionary Strike Group 2
Commander, Expeditionary Strike Group 3
Commander, Expeditionary Strike Group 7/Task Force 76
Commander, First Naval Construction Division; Commander, Naval Construction Forces Command
Commander, Fleet and Industrial Supply Centers
Commander, Fleet Readiness Centers, CNAF
Commander, Joint Task Force, Guantanamo
Commander, Logistics Group, Western Pacific; Commander, Task Force 73
Commander, Mid-Atlantic Regional Maintenance Center
Commander, Military Sealift Command
Commander, National Naval Medical Center; Chief, Navy Medical Corps
Commander, Naval Air Force, U.S. Atlantic Fleet
Commander, Naval Air Warfare Center Aircraft Division; Assistant Commander for Research and Engineering, Naval Air Systems Command
Commander, Naval Air Warfare Center Weapons Division, China Lake and Point Mugu, Calif.
Commander, Naval Education and Training Command
Commander, Naval Expeditionary Logistics Support Force
Commander, Naval Facilities Engineering Command, Atlantic Division
Commander, Naval Facilities Engineering Command, Pacific
Commander, Naval Facilities Engineering Command; Chief of Civil Engineers
Commander, Naval Inventory Control Point
Commander, Naval Meteorology and Oceanography Command
Commander, Naval Mine and Anti-Submarine Warfare Command
Commander, Naval Personnel Development Command
Commander, Naval Region Northwest

Commander, Naval Safety Center
Commander, Naval Service Training Command
Commander, Naval Special Warfare Command
Commander, Naval Strike and Air Warfare Center
Commander, Naval Supply Systems Command and Chief of Supply Corps
Commander, Naval Surface Force, U.S. Atlantic Fleet
Commander, Naval Surface Warfare Center; Deputy Commander for Warfare Systems Engineering, Naval Sea Systems Command
Commander, Naval Undersea Warfare Center; Deputy Commander for Undersea Warfare and Undersea Technology, Naval Sea Systems Command
Commander, Naval Undersea Warfare Center; Director, Undersea Technology; Naval Sea Systems Command (Sea 073)
Commander, Navy Exchange Service Command, Fleet and Family Support Services
Commander, Navy Expeditionary Combat Command
Commander, Navy Medicine East; Commander, Naval Medical Center Portsmouth
Commander, Navy Medicine Support Command; Chief, Navy Dental Corps
Commander, Navy Medicine West, Naval Medical Center San Diego
Commander, Navy Personnel Command and Deputy Chief of Naval Personnel
Commander, Navy Recruiting Command
Commander, Navy Region Europe; Commander, Maritime Air Naples
Commander, Navy Region Hawaii; Commander, Naval Surface Group Middle Pacific
Commander, Navy Region Mid-Atlantic
Commander, Navy Region Midwest
Commander, Navy Region Southeast
Commander, Navy Region Southwest
Commander, Navy Reserve Forces Command

Commander, Navy Reserve Intelligence Command
Commander, Navy Warfare Development Command
Commander, Operational Test and Evaluation Force
Commander, Pacific Division, Naval Facilities Engineering Command
Commander, Patrol and Reconnaissance Force Seventh Fleet; Commander, Patrol and Reconnaissance Force Fifth Fleet; Commander, Fleet Air Western Pacific
Commander, Patrol and Reconnaissance Group
Commander, Space and Naval Warfare Systems Command
Commander, Strike Force Training Atlantic
Commander, Strike Force Training Pacific
Commander, Submarine Force, U.S. Pacific Fleet
Commander, Submarine Group 2
Commander, Submarine Group 7; Commander, Task Force 74; Commander, Task Force 54
Commander, Submarine Group Trident
Commander, U.S. Naval Forces Japan
Commander, U.S. Naval Forces Korea
Commander, United States Naval Forces Southern Command
Commanding Officer, Navy Air Logistics Office
Commanding Officer, Navy Reserve Commander, Naval Forces Europe/Commander 6th Fleet Detachment 802
Commanding Officer, Navy Supply Support Battalion 1
Commanding Officer, USS George Washington (CVN 73)
Communications Division Chief, Strategic Effects, MNF-I; Embassy Annex, Baghdad
Deputy Assistant Secretary of Defense (Joint Communication)
Deputy Assistant Secretary of the Navy for Acquisition and Logistics Management
Deputy Assistant Secretary of the Navy for Acquisition Management for the Assistant Secretary of the Navy (Research, Development and Acquisition)
Deputy Assistant Secretary of the Navy, International Programs Director, Navy International Programs Office
Deputy Chief for Operations, Bureau of Medicine and Surgery

Deputy Chief of Chaplains for Reserve Matters; Director of Religious Programs, Marine Force Reserve
Deputy Chief of Chaplains; Chaplain of the Marine Corps
Deputy Chief of Civil Engineers and Deputy Commander for Contingency Engineering
Deputy Chief of Information
Deputy Chief of Staff for Logistics, Fleet Supply and Ordnance, U.S. Pacific Fleet
Deputy Chief of Staff for Operations, Allied Maritime Component Command, Northwood, UK
Deputy Chief of Staff for Operations, Training and Readiness, U.S. Pacific Fleet
Deputy Chief of Staff for Plans, Policies and Requirements, Commander, Pacific Fleet (N5/N8)
Deputy Command Surgeon, U.S. Fleet Forces Command; Command Surgeon, U.S. Joint Forces Command; Medical Advisor, Supreme Allied Command for Transformation
Deputy Commander Force Integration, National Capital Area; Deputy Director of the Nurse Corps
Deputy Commander for Mobilization, Naval Sea Systems Command, Space and Naval Warfare Systems Command
Deputy Commander for Ship Design, Integration and Engineering, Naval Sea Systems Command
Deputy Commander for Total Force Integration, Navy Medicine Support Command; Chief, Reserve Medical Corps, Bureau of Medicine and Surgery
Deputy Commander for Undersea Warfare (NAVSEA 07), Naval Sea Systems Command
Deputy Commander for Warfare Systems Engineering, SEA-06, Naval Sea Systems Command
Deputy Commander Submarine Force, U.S. Pacific Fleet; Deputy Commander, Navy Region Hawaii
Deputy Commander, Combined Joint Task Force, Horn of Africa
Deputy Commander, First Naval Construction Division
Deputy Commander, Joint Functional Component Command for Space and Global Strike

Deputy Commander, Logistics, Maintenance and Industrial Operations, Naval Sea Systems Command
Deputy Commander, Naval Inventory Control Point in Mechanicsburg, Pa.
Deputy Commander, Naval Operational Logistics Support Center
Deputy Commander, Navy Expeditionary Combat Command
Deputy Commander, Navy Region Southeast
Deputy Commander, Navy Region Southwest
Deputy Commander, Second Fleet
Deputy Commander, Third Fleet
Deputy Commander, U.S. Naval Forces, U.S. Central Command
Deputy Commander, U.S. Seventh Fleet
Deputy Commander/Chief of Staff, U.S. Pacific Fleet
Deputy Director for Operations, J3, Joint Staff
Deputy Director for Strategy and Policy, J-5, Joint Staff
Deputy Director of Customer Relationships, Signals Intelligence Directorate, National Security Agency
Deputy Director of Operations, NMCC, JCS
Deputy Director of Operations, United States Transportation Command
Deputy Director of Surface Warfare (CNO N86B)
Deputy Director, Air Warfare (OPNAV N88B/N882)
Deputy Director, Division of Submarine Warfare (OPNAV N77B)
Deputy Director, Expeditionary Warfare (N85B)
Deputy Director, Fleet Readiness Division (OPNAV N43B)
Deputy Director, Naval Medicine, Office of the Chief of Naval Operations
Deputy Director, Standing Joint Task Force Headquarters–North, U.S. Northern Command
Deputy Director, Strategy, Plans and Policy, U.S. Central Command
Deputy Director, U.S. Strategic Command Center for Combating Weapons of Mass Destruction
Deputy for Combat Systems/Weapons (N86F)

Deputy Judge Advocate General; Commander, Naval Legal Service Command
Deputy Regional Chaplain, Navy Region Mid-Atlantic
Deputy Surgeon General
Deputy to the Director J5, United States Central Command
Deputy Vice Commander, Naval Air Systems Command; Director, Naval Air Reserve Program
Deputy, Joint Forces Maritime Component Commander, U.S. Third Fleet
Deputy, The Medical Officer of the Marine Corps
Direct Reporting Program Manager for Navy Marine Corps Intranet (DRPM (NMCI))
Director for Intelligence (J2), Joint Staff
Director for Intelligence, United States Pacific Command
Director for Logistics and Engineering, U.S. Northern Command
Director for Manpower and Personnel, J1, The Joint Staff
Director for Operational Plans and Joint Force Development, Joint Staff
Director for Strategic Capabilities Policy Directorate of Defense Policy and Strategy, National Security Council
Director for Strategy and Policy (N5SP)
Director of Global Operations, Naval Network Warfare Command
Director of Information, Plans, and Security OPNAV
Director of Intelligence
Director of Logistics and Security Assistance (ECJ4)
Director of Operations (J3), United States Navy; United States European Command
Director of the Navy Nurse Corps and the Chief of Staff, Bureau of Medicine and Surgery
Director, Air Warfare Division (OPNAV N88)
Director, Assessment Division/Capability Analysis Group (OPNAV N81/N00X)
Director, Aviation and Aircraft Carrier Plans and Requirements
Director, CNO Environmental Readiness Division (N45)

Director, Command Control Systems, NORAD-USNORTHCOM J6
Director, Communications Systems Acquisition and Operations Directorate, National Reconnaissance Office
Director, Fleet Maintenance; Commander, U.S. Fleet Forces Command (N43)
Director, Fleet Readiness Division (OPNAV N43)
Director, Health Services/The Medical Officer of the Marine Corps
Director, Information Operations (OPNAV N3IO); Deputy Director for Cryptology Division (OPNAV N2C)
Director, Joint Innovation & Experimentation, United States Joint Forces Command
Director, Joint Public Affairs Support Element–Reserve
Director, Joint Reserve Forces
Director, Maritime Partnership Programs; Commander, U.S. Naval Forces Europe; Commander, Sixth Fleet
Director, Medical Service Corps; Deputy Chief of Staff, Human Resources, Bureau of Medicine and Surgery
Director, Military Personnel Plans and Policy Division (N13)
Director, Naval Forces Europe/Sixth Fleet Plans and Operations; Deputy Commander, Sixth Fleet; Commander, Submarines, Allied Naval Forces South; Commander, Submarine Group Eight
Director, Navy Medical Service Corps, Bureau of Medicine and Surgery
Director, Office of Budget; Office of the Assistant SECNAV for Financial Management and Comptroller; Director, Fiscal Management Division, N82
Director, Operations Division, FMB1/N821
Director, Programming Division (OPNAV N80)
Director, Space and Network Warfare Program
Director, Standing Joint Force Headquarters
Director, Strategic Plans and Policy
Director, Strategic Systems Programs

Director, Strategy & Policy, Resources and Transformation, U.S. Naval Forces Europe
Director, Strategy, Policy, Programs, and Logistics Directorate (J5/4), USTRANSCOM
Director, Submarine Warfare Division (OPNAV N87)
Director, Warfare Integration (N6F)
Director, Warfare Integration/Senior National Representative (N8F)
Director, White House Military Office
Executive Officer, Navy Reserve Naval Special Warfare Command
Judge Advocate General of the Navy
Maritime BMD, Commander Second Fleet
Navy Medicine West, Deputy for Total Force Integration and Deputy Chief of the Navy Dental Corps
Oceanographer/Navigator of the Navy
President, Board of Inspection and Survey
President, Naval War College
Principal Director for Operations and Deputy Commander, Joint Task Force–Global Network Operations (JTF-GNO)
Program Director, Aegis Ballistic Missile Defense, Missile Defense Agency
Program Executive Officer for Enterprise Information Systems (PEO-EIS)
Program Executive Officer for Integrated Warfare Systems
Program Executive Officer for Submarines
Program Executive Officer, Joint Strike Fighter Program
Program Executive Officer, Ships
Program Executive Officer, Strike Weapons and Unmanned Aviation, Naval Air Systems Command
Program Executive Officer, Tactical Aircraft Programs
Regional Director, TRICARE Regional Office West
Representative, Senior Interagency Strategy Team, National Counterterrorism Center

Reserve Affairs, Navy Medicine East, Associate Chief, Bureau of Medicine and Surgery, Health Care Operations and Deputy Director for Reserve Affairs, Medical Service Corps
Reserve Deputy and Chief of Staff, U.S. Pacific Fleet
Reserve Deputy Commander; Commander, Navy Installations Command
Reserve Deputy, Warfare Requirements and Programs (OPNAV N6/7)
Senior Advisor to the Assistant Deputy Surgeon General for Total Force Integration
Senior Fellow, CNO Strategic Studies Group
Senior Military Assistant to the Secretary of the Navy
Special Assistant to the Commander, U.S. Central Command
Standing NATO Maritime Group 2
Surgeon, U.S. Pacific Command
U.S. Pacific Defense Representative to Guam, Commonwealth of the Northern Marianas, Federated States of Micronesia, and Republic of Palau
Vice Commander, Naval Air Forces; Commander, Naval Air Force Reserve
Vice Commander, Naval Network Warfare Command
Vice Commander, Naval Reserve Forces Command; Deputy Commander, Navy Region Northwest
Vice Commander, Naval Submarine Forces
Vice Commander, Naval Surface Forces
Vice Commander, Navy Expeditionary Combat Command; Commander, Navy Expeditionary Logistics Support Group
Vice Commander, Space and Naval Warfare Systems Command
Vice Commander, U.S. Fleet Forces Command
Vice Director for Operations, J-3, The Joint Staff

APPENDIX B

Definitions of Domain Expertise

Acquisition Management: Skill in directing government-funded programs that deliver new, improved, or continuing materiel, weapon, or information systems or service capability providing effective, affordable, and timely systems to users.

Acquisition Professional: Defense Acquisition Workforce Improvement Act (DAWIA) Certification as an acquisition professional.

Aerospace Engineering and Maintenance: Skill in managing all aspects of the life cycle of naval aviation programs/weapon systems and space-based sensors/weapon systems.

Air Warfare: Skill in and planning and executing air defense and air interdiction in support of air superiority.

Civil Engineering: Skill in executing the planning, design, construction, operation, and maintenance of shore facilities as well as contingency engineering, environmental, and natural resource management, ocean and littoral engineering, and naval architecture programs.

Counterterrorism: Skill in planning and conducting operations that include the offensive measures taken to prevent, deter, preempt, and respond to terrorism. Also called CT.

Cryptology: Skill in the science of decoding and/or encoding information for hidden, disguised, or encrypted communications. It includes communications security and communications intelligence.

Expeditionary Warfare: The overall ability to organize an armed force to accomplish a specific warfare objective in a foreign country from the sea by a deployed self-sustaining force.

Financial Management: Skill in planning and integrating timely, accurate, and useful information for policies, procedures, and direction on accounting, finance, management control, financial services, and financial systems resources.

Human Resources: Skill in exercising oversight and responsibilities of employment of both military personnel and civilians, focusing on compensation, employee relations, benefits, training, developing and implementing policies and procedures, management consulting, employee counseling, and safety.

Information Professional: Skill in managing, planning, acquiring, and integrating naval and expeditionary networks, command and control systems, and knowledge management processes.

Information Warfare: Skill in planning and executing actions to achieve information superiority by affecting adversary information, information-based processes, and information systems (both human and machine) while defending one's own similar resources.

Installation Management: Skill in the exercise of executive, administrative, and supervisory direction and oversight over subordinate regions and/or installations for the delivery/provision of day to day installation services.

Intelligence: Skill in planning, collecting, analyzing, producing, and disseminating intelligence to inform policy and support operations.

Joint and Combined Warfare: Skill in planning and executing operations in support of joint or coalition objectives.

Logistics and Readiness: Skill in planning and executing the projection, movement and sustainment, reconstitution, and redeployment of operating forces and systems.

Medical Administration: The ability of the leader to keep detailed notes on patients in the field of operations/and or medical facilities. Leaders must be able to administratively keep patient records in order, appointment scheduling, accounts payable and receivables, and must interface with Tricare and other DoD healthcare agencies to ensure proper care of patients.

Mine and Undersea Warfare: Skill in planning and executing operations to support the control of the battlespace through offensive or defensive mining operations. Also called MIW.

Nuclear Propulsion: Skill in highly technical aspects of program management; research and engineering; and test and evaluation to support supervising, operating, and maintaining nuclear propulsion plants aboard submarines and surface ships.

Oceanography: Skill in collecting, interpreting, and applying meteorological and oceanographic data to support safety at sea; strategic and tactical warfare; and weapon system design, development, and deployment.

Operational Medicine: The ability of the leader to provide professional and technical support and consultative services in operationally related Fleet and Fleet Marine Force medical matters worldwide.

Operations Analysis: Skill in designing and analyzing research systems to optimize resource allocation and improve decisionmaking.

Public Affairs: Skill in providing strategic counsel, and operational planning and tactical execution of communications.

Reserve Integration: The ability of the leader to ensure that active and reserve components will be broadly and seamlessly integrated, and indivisible as a balanced warfighting force.

Ship Engineering and Repair: Skill in managing all aspects of Fleet maintenance: repair, modernization, and disposal with core areas of expertise in program management; Nuclear Engineering; naval architecture; Hull, Mechanical, and Electrical (H,M&E) systems; and industrial and corporate operations.

Space Warfare: The ability to use all offensive measures to hit the defined enemy target and to defensively use measures to destroy attacking enemy vehicles (including missiles) while in space, or to nullify or reduce the effectiveness of such attack.

Special Warfare: Skill in planning and executing warfare in hostile, denied, or politically sensitive environments to achieve military, diplomatic, informational, and/or economic objectives employing military capabilities for which there is no broad conventional force requirement.

Strategic Plans and Policy: Skill in developing and evaluating the application of strategy, doctrine, and concepts.

Submarine Warfare: Skill in planning and executing tactical and strategic operations for tracking and destroying enemy submarine and surface assets while remaining undetected, and gaining and sustaining assured access for military assets in denied areas while denying the enemy to do the same.

Supply Management: Skill in procuring, producing, and delivering products and services to the military and other DoD customers.

Surface Warfare: Skill in planning and executing operations designed to project power over land, protect naval forces operating at sea, and provide access to maritime theaters in support of maritime superiority. Also called SUW.

UCMJ/Legal: The ability of a leader to understand and utilize the Congressional Code of Military Criminal Law applicable to all military members worldwide.

Unmanned Warfare: The ability to demonstrate the technical feasibility, military utility, and operational value for a networked system of high performance, weaponized unmanned air/sea vehicles to effectively and affordably prosecute 21st-century combat missions, including Suppression of Enemy Air or Sea Defenses, using surveillance and precision strike within the emerging global command and control architecture.

Warfare Resources Management: Skill in planning and executing operations to ensure the efficient and effective use of available platforms, personnel, and budgets.

APPENDIX C

Cross-Functional Expertise

Leading Change

1. Skill in creating, communicating, and executing a strategic vision that will impact the future of the Navy and the Nation.
2. Skill in linking innovative and strategic thinking in my command or organization to Navy and National strategies.
3. Skill in demonstrating external awareness of world and national affairs that impact the Navy's strategic vision.
4. Skill in exercising flexibility to stimulate process development, evaluate new ideas, and achieve Navy Vision.
5. Skill in providing clear guidance on expectations, achieving results, risk management, and mission accomplishment.
6. Ability to engage in continual learning opportunities to master new knowledge, pursue self-development, and grasp new information.

Leading People

1. Skill in motivating, inspiring, and mentoring military personnel through a positive attitude, enthusiastic leadership, and ethical behavior.
2. Skill in motivating, inspiring, and mentoring civilian personnel (DoD and contractors) through a positive attitude, enthusiastic leadership, and ethical behavior.

3. Skill in leading by professional example to promote team building and personnel development.
4. Skill in managing conflict in a crisis by identifying potential situations that could result in unpleasant confrontations.
5. Skill in managing conflict in a combat or wartime situation to maximize force effectiveness and enhance mission accomplishment.
6. Skill in leveraging an ethnically and culturally diverse workforce to improve working environment and capitalize on achievements of each individual.

Stewarding Resources

1. Comprehensive knowledge of and effective use of the Planning, Programming, and Budgeting System (PPBS).
2. Proficiency in leveraging technology to enhance business acumen and skills, developing assessment processes, and analyzing alternatives.
3. Broad understanding of principals of financial management and marketing sufficient to ensure appropriate funding and prioritization.
4. Ability to assess current and future human resources and staffing requirements based on organizational goals and budget realities.
5. Technical skill necessary to understand and apply procedures, requirements, regulations and policies, and to make sound resource decisions.
6. Analytical and research abilities to frame problems, synthesize issues, formulate solutions, and recommend courses of action.

Externally Networking

1. Skill in influencing and negotiating with people at all levels including civic leaders, Joint Staff, OSD, Inter-Agency, Congress, and White House.
2. Diplomacy, political awareness, international savvy, and negotiating skills necessary to partner with foreign navies and governments.
3. Interpersonal skills necessary to communicate extemporaneously to build networks and coalitions, and to accomplish missions.
4. Skill in preparing and delivering quality oral presentations and written communications to demonstrate expertise and persuade others to accomplish objectives.
5. Capacity to build and sustain effective networks through use of information technology.
6. Knowledge of how and who to ask for the capabilities of other Services and Agencies as part of joint operations and warfare.

Integrating Results

1. Skill in employing force to achieve Joint, Coalition, and Interagency objectives and missions.
2. Knowledge of command and control, roles, doctrines, missions, and capabilities of Joint and Coalition Forces to conduct operational planning and execution in a complex environment.
3. Resilience and flexibility to deal effectively with change, to focus on objectives under pressure, and to recover quickly from setbacks.
4. Ability to exercise good judgment, perception, adaptiveness, and common sense necessary to integrate priorities and eliminate irrelevant information.
5. Skill in measuring readiness and operational effectiveness to achieve and sustain Joint Operational Excellence.

6. Ability to effectively advocate the use of Naval Forces and sea power within and outside the Navy and Marine Corps.
7. Understanding the structures and organizations of other Services so that you can grasp how or why joint decisions might be made.
8. Ability to integrate practices and rules of each Service to overcome cultural and operational differences and achieve joint objectives.

Accomplishing Mission

1. Ability to exercise responsibility, good judgment, authority, and accountability in all aspects of this billet.
2. Ability to develop and maintain effective controls which ensure the integrity of the command/organization while holding yourself and others accountable for rules and regulations.
3. Proficiency in problem solving and continuous improvement techniques and processes to achieve concise and powerful results.
4. Ability to create operational and work environments where decisiveness and risk management will optimize outcomes and force effectiveness.
5. Skill in fully integrating naval forces into the joint team to maximize our advantages with dominant, precise, and persistent power.
6. Ability to provide combatant commanders with flexible, agile, and capable naval forces for today's dynamic and uncertain strategic environment.

APPENDIX D

Survey Screenshots and Additional Definitions Used in the Survey

Our survey of Navy flag officers was administered via computer. This appendix presents the screenshots that respondents saw during the survey.

Definitions Used During the Survey

During the survey, important terms were highlighted on the screen. Respondents could view the definitions of these terms by hovering the mouse over the words on the screen. Many of these were the the definitions of each area of expertise, which are presented in Appendix B. Other definitions available during the survey are presented below.

Domain Knowledge: Subject matter expertise; developed knowledge and skill in a specific field of interest or content. Some U.S. Navy examples are Surface Warfare, Acquisition Management, Financial Management, and Human Resources.

Enterprise Knowledge: Expertise with regard to how the U.S. Navy functions, including an understanding of the functioning of U.S. Navy organizations and other organizations outside of the U.S. Navy that are relevant to the billet. A U.S. Navy example is: Understanding the structures, organizations, capabilities, and cultures of the other military services.

Job Book: The Job Book contains a summary of the competencies identified as critical for successful performance in Navy flag and SES billets.

Leadership Skill: Skill in guiding, directing, or influencing people. A U.S. Navy example is: Motivating, inspiring, and mentoring military personnel.

Management Skill: Skill in organizing and controlling the affairs of an enterprise or a particular sector of an enterprise. A U.S. Navy example is: Skill in applying procedures, requirements, regulations, policies, and business principles to make sound resource decisions.

Categories of Domain Knowledge

We further categorize "Domain Knowledge" in terms of the depth of knowledge, skill, or experience that is required.

Primary Expertise: A collection of critical occupational skills required for successful performance in the billet. The designation "Primary Expertise" signifies that the skills are required at a depth of expertise that can only be attained through a career-long focus. Marking more than one "P" for a billet signifies that an individual with expertise in any one of the occupational areas marked "P" could be equally successful in this billet.

Secondary Expertise: A collection of critical occupational skills required for successful performance in the billet. The designation "Secondary Experience" signifies that the skills are required at a depth of experience that can be attained through one or two tours of experience and relevant education or training. Marking more than one "S" for a billet signifies that an individual with experience in any one of the occupational areas marked "S" could be equally successful in this billet.

Familiarity: A collection of critical occupational knowledge required for successful performance in the billet. The designation "Familiarity" signifies that the knowledge is required at a depth of understanding that a well informed customer would have. This level of knowledge might be attained through education or training alone. A billet will typically require familiarity with many occupational areas at this level.

Any Primary Skill: No specific primary-level domain knowledge is required for success in this billet. (Equivalent to saying that a flag officer from any community could perform successfully in this billet.)

Any Secondary Skill: No specific secondary-level domain knowledge is required for success in this billet.

Survey Screen Shots

The remainder of this appendix presents the screenshots from the survey.

Figure D.1
Survey Screen 1: Welcome Page

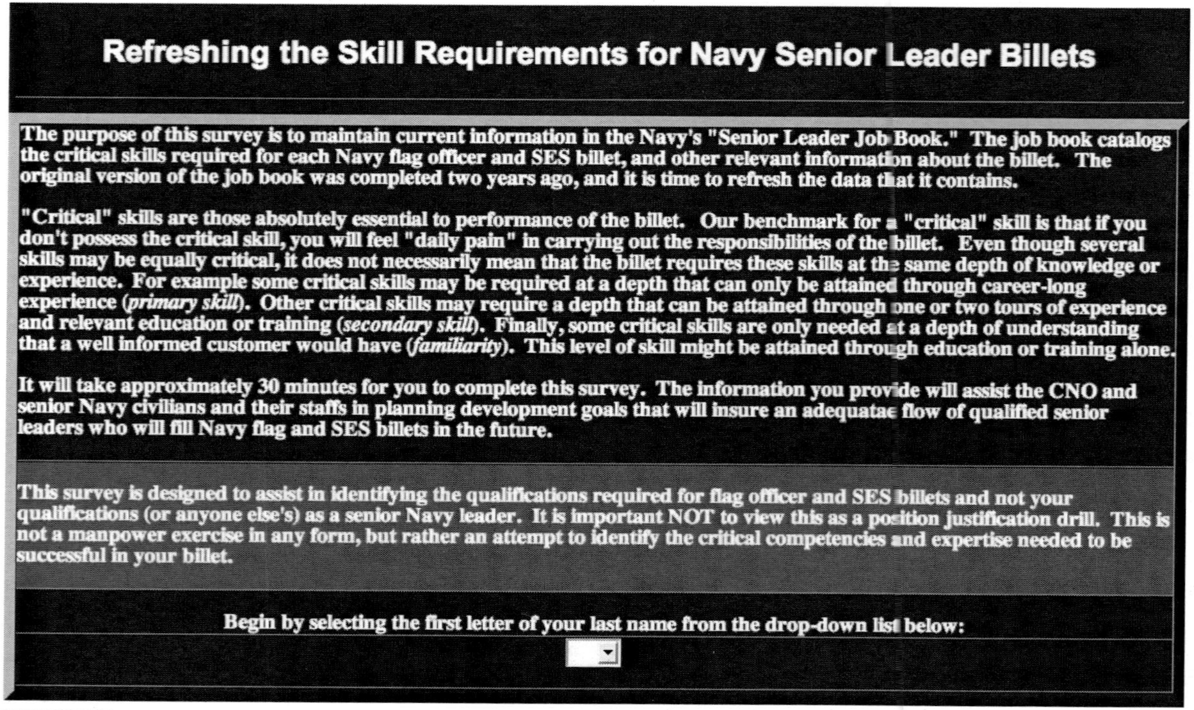

RAND *MG618-D.1*

Figure D.2
Survey Screen 2: Introduction

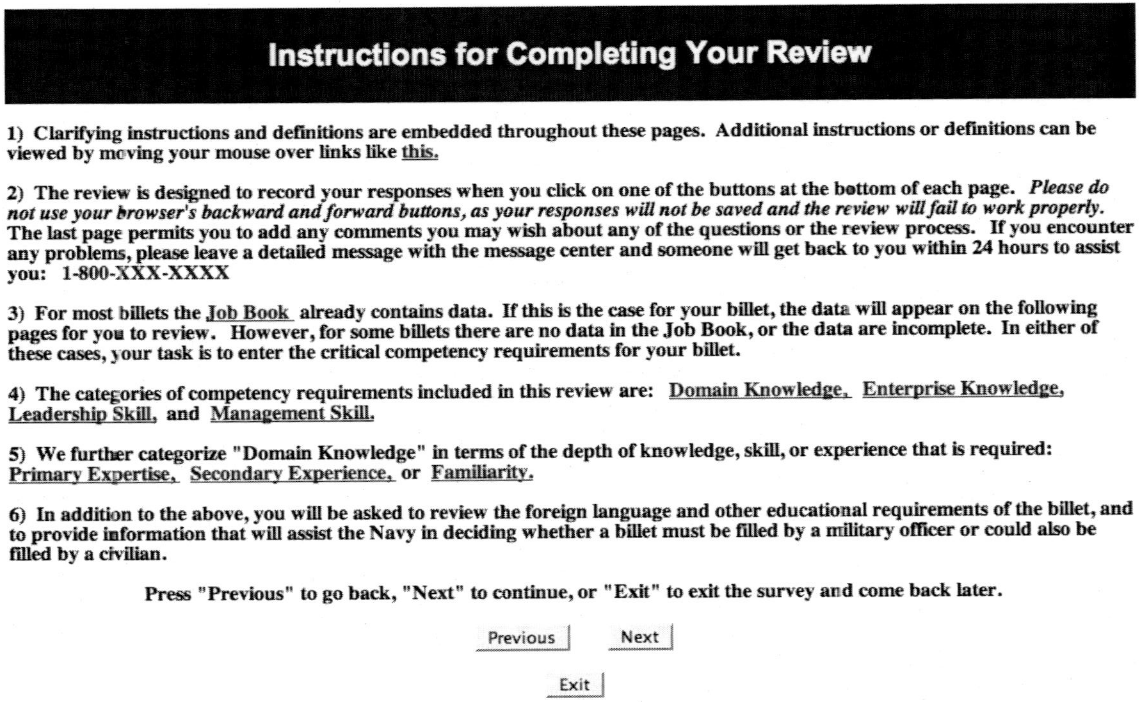

RAND *MG618-D.2*

Figure D.3
Survey Screen 3: Domain Knowledge Requirements

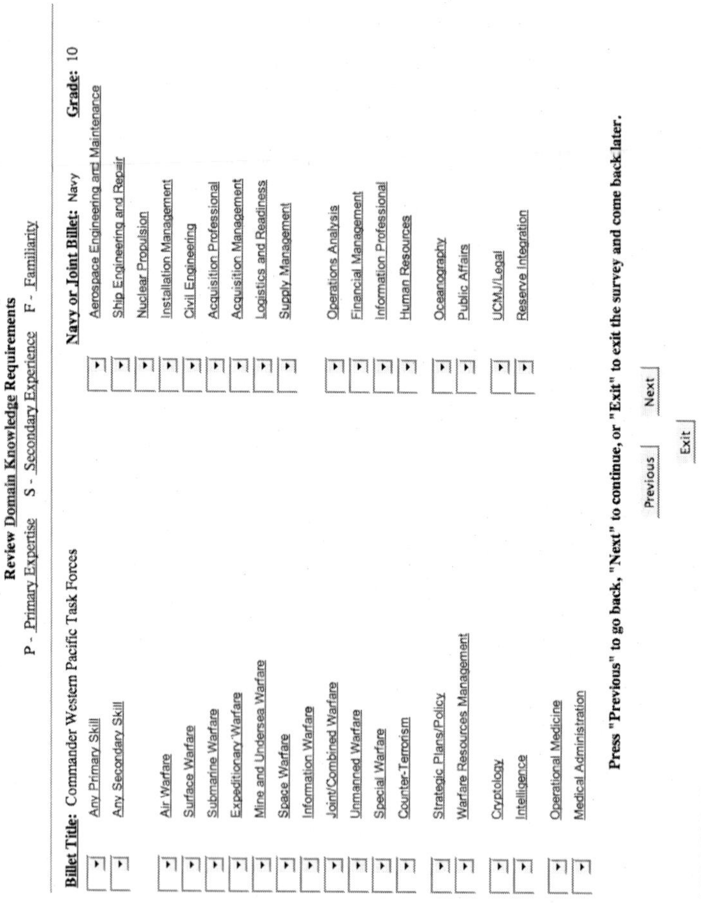

RAND *MG618-D.3*

Figure D.4
Survey Screen 4: Enterprise Knowledge, Leadership, and Management Skill Requirements (1 of 2)

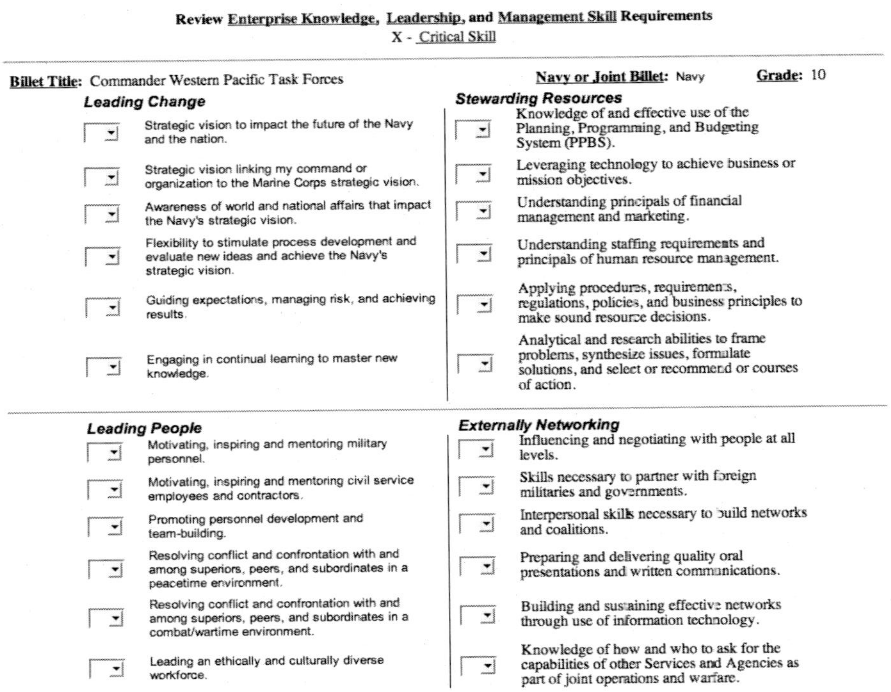

RAND *MG618-D.4*

Figure D.5
Survey Screen 5: Enterprise Knowledge, Leadership, and Management Skill Requirements (2 of 2)

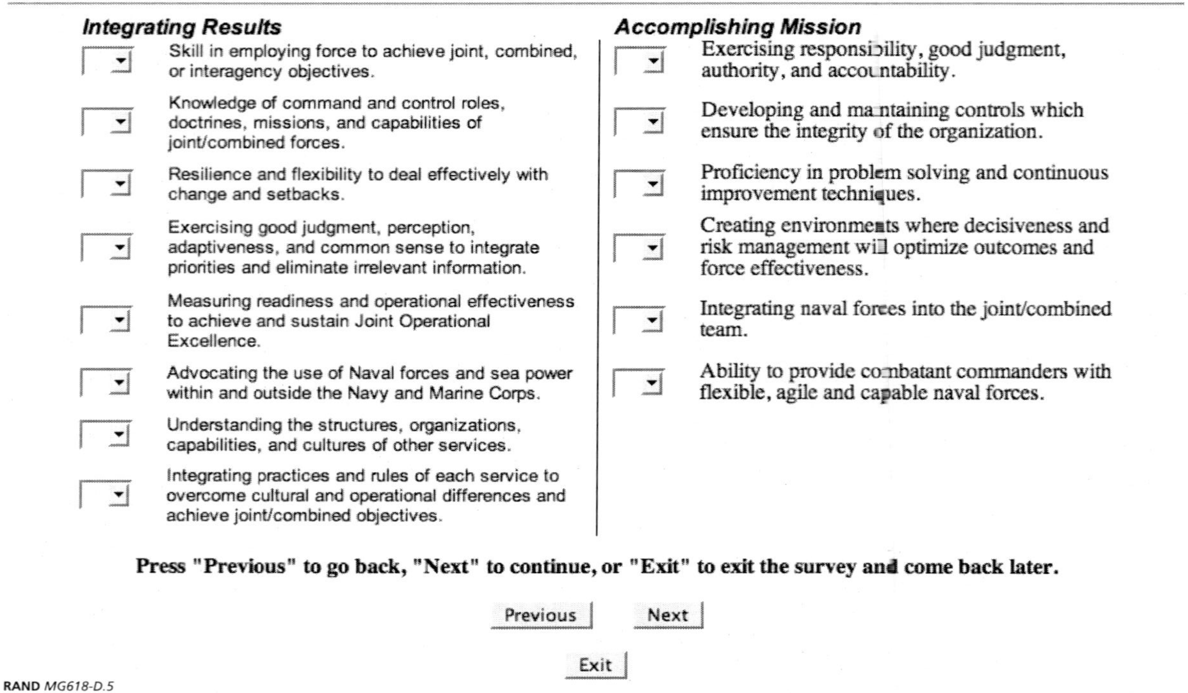

Integrating Results

- Skill in employing force to achieve joint, combined, or interagency objectives.
- Knowledge of command and control roles, doctrines, missions, and capabilities of joint/combined forces.
- Resilience and flexibility to deal effectively with change and setbacks.
- Exercising good judgment, perception, adaptiveness, and common sense to integrate priorities and eliminate irrelevant information.
- Measuring readiness and operational effectiveness to achieve and sustain Joint Operational Excellence.
- Advocating the use of Naval forces and sea power within and outside the Navy and Marine Corps.
- Understanding the structures, organizations, capabilities, and cultures of other services.
- Integrating practices and rules of each service to overcome cultural and operational differences and achieve joint/combined objectives.

Accomplishing Mission

- Exercising responsibility, good judgment, authority, and accountability.
- Developing and maintaining controls which ensure the integrity of the organization.
- Proficiency in problem solving and continuous improvement techniques.
- Creating environments where decisiveness and risk management will optimize outcomes and force effectiveness.
- Integrating naval forces into the joint/combined team.
- Ability to provide combatant commanders with flexible, agile and capable naval forces.

Press "Previous" to go back, "Next" to continue, or "Exit" to exit the survey and come back later.

Previous Next

Exit

RAND *MG618-D.5*

Figure D.6
Survey Screen 6: Education Requirements

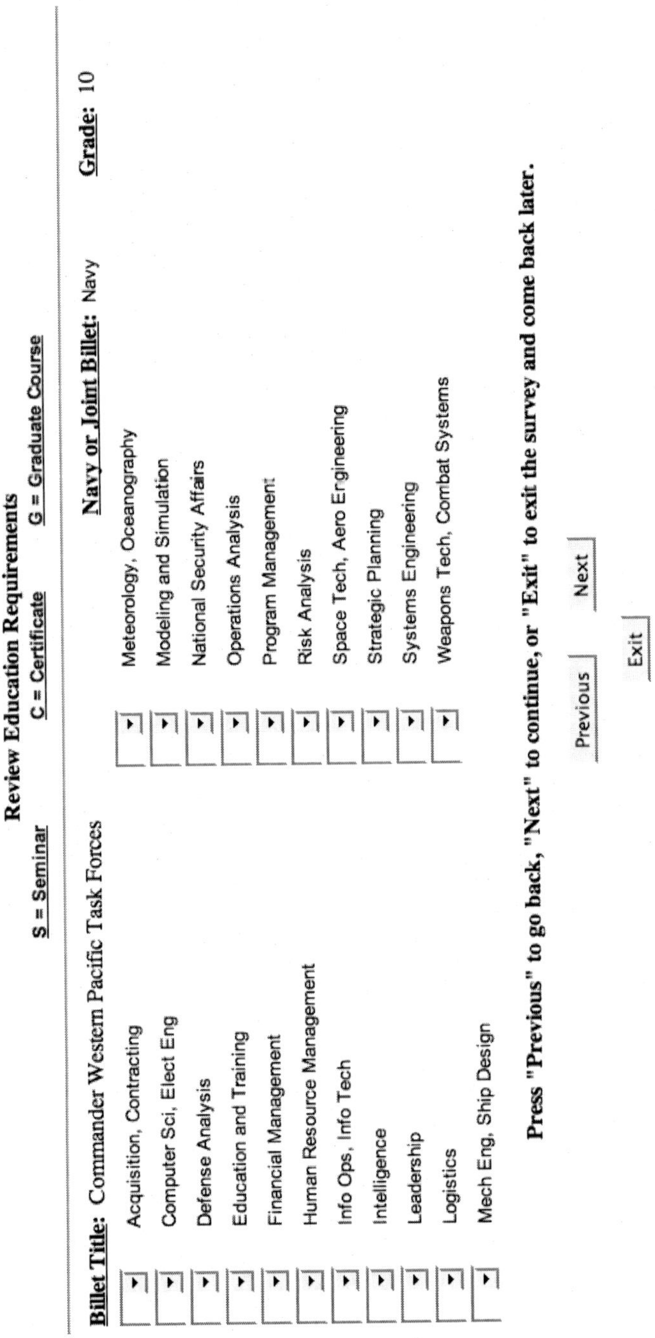

RAND *MG618-D.6*

Figure D.7
Survey Screen 7: Foreign Language Requirements

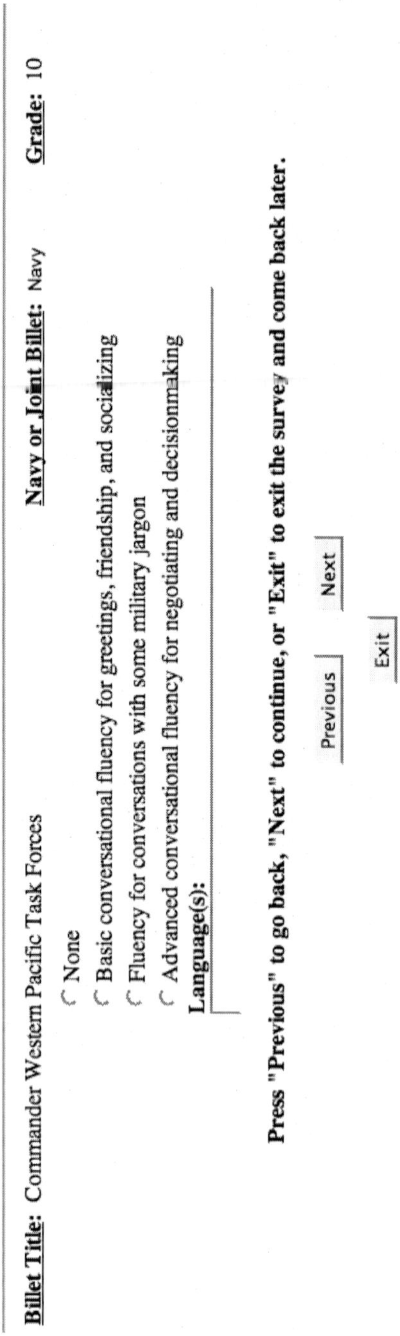

RAND *MG618-D.7*

Figure D.8
Survey Screen 8: Military/Civilian Fill Requirements

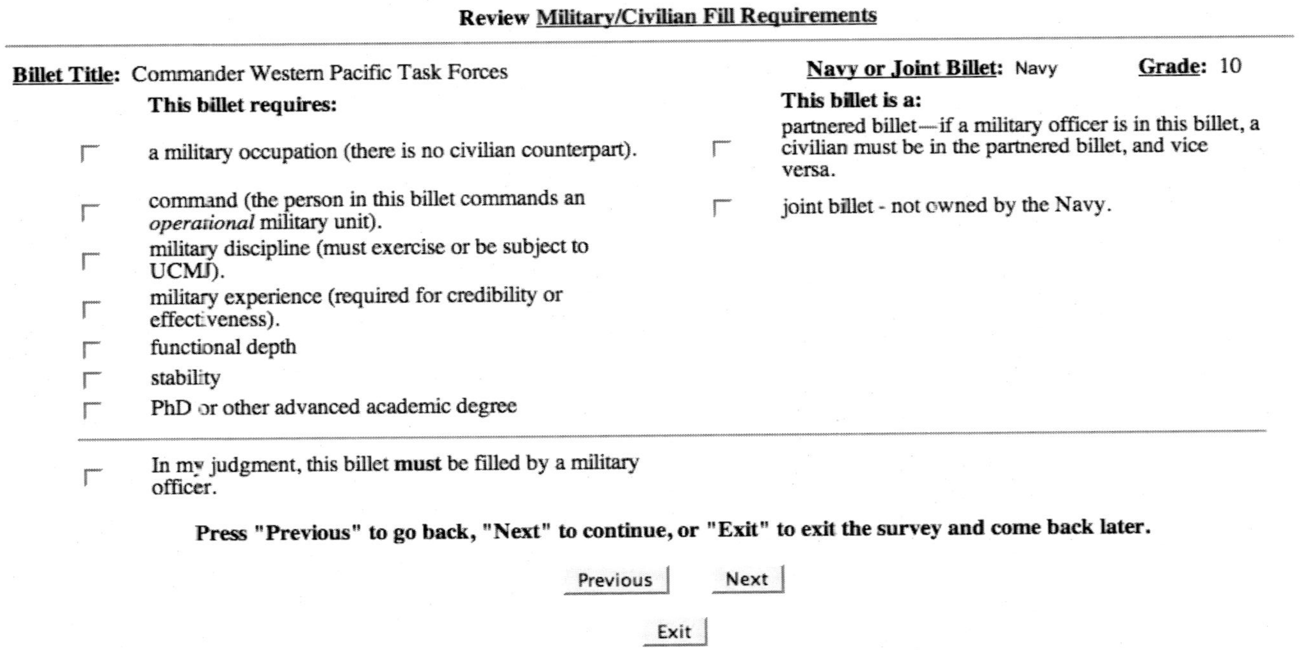
Review Military/Civilian Fill Requirements

Billet Title: Commander Western Pacific Task Forces

Navy or Joint Billet: Navy **Grade:** 10

This billet requires:

- a military occupation (there is no civilian counterpart).
- command (the person in this billet commands an *operational* military unit).
- military discipline (must exercise or be subject to UCMJ).
- military experience (required for credibility or effectiveness).
- functional depth
- stability
- PhD or other advanced academic degree

This billet is a:

- partnered billet—if a military officer is in this billet, a civilian must be in the partnered billet, and vice versa.
- joint billet - not owned by the Navy.

- In my judgment, this billet **must** be filled by a military officer.

Press "Previous" to go back, "Next" to continue, or "Exit" to exit the survey and come back later.

Previous Next

Exit

RAND *MG618-D.8*

Figure D.9
Survey Screen 9: Comments

Comments

(Commander Western Pacific Task Forces)

This is the last page of the review. Thank you for taking the time to respond thoughtfully.

Please use the space below for any further comments you wish to provide about the skills and experiences required for successful performance in this position.

Be sure to click on one of the buttons below the comment box to record your comments and to notify us that you are done with your review of this position.

Previous Question

I am finished reviewing the billet listed above

RAND *MG618-D.9*

APPENDIX E

Formulation of the Mathematical Program to Determine Average Flows of Inventory Through the Flag Officer Ranks

This appendix contains a more detailed explanation and corresponding mathematical expressions of the computations for modeling the average flows of inventory in the flag officer ranks outlined in Chapter Five.

Physics of Inventory-to-Billet Flows

The model is concerned with long-term steady-state averages. One set of variables describes inventory. Inventory variables specify the steady-state average number of flag officers at a grade who possess a specific combination of primary and secondary areas of expertise and grade. In our equations we use *Inventory*(c,g), where c is the combination of areas of expertise and g is the grade.

Companion variables are the average annual promotions of officers having particular combinations of primary and secondary areas of expertise into the various grades. We indicate these variables by *Promotions*(c,g).

From past experience, we have data on the average time an officer is at each grade. (The average times are around three years.) We denote these by *TIG*(g).

With knowledge of average times in grade, average inventory and promotion rate variables are related by Little's Theorem (Little, 1961, pp. 383–387), which asserts that for a particular combination of primary and secondary areas of competency at a grade,

$$Inventory(c,g) = Promotions(c,g) \times TIG(g). \tag{1}$$

Matching Inventory to Billets

The second key ingredient in the model's logic is connecting inventory with billets. This is done with variables that specify how much inventory associated with primary and secondary areas of expertise at a grade fills each billet. The variables are denoted by *Assign*(c,j), where j indicates a specific billet (job). *Assign* variables are involved in both "supply" and "demand" relationships. Supply refers to how much of the various kinds of inventory are used, and demand refers to how billets are filled.

A supply relationship says that for a combination of primary and secondary areas of expertise, the sum of *Assign* variables taken over jobs that can use the expertise combination should equal the *Inventory* for the expertise combination and grade. The formal equation is

$$Inventory(c,g) = \sum_{j \in J(g)} Assign(c,j) \ , \tag{2}$$

where $J(g)$ is the set of billets at grade g. This equation is not quite complete. There is information about which areas of expertise are allowed for each job, such that inappropriate (c,j) combinations can be eliminated from the sum on the right-hand side.

A demand relationship asserts that, for each billet, the sum of *Assign* variables over combinations of expertise that could be used to fill the billet must equal the fraction of time that a Navy officer would be expected to occupy the billet. We call that fraction *Requirement*(j). It is usually equal to one, but could be less for joint billets that are rotated among the services.

A demand equation for a billet is

$$\sum_{c} Assign(c,j) = Requirement(j). \qquad (3)$$

Figure E.1 depicts how the three equations interact. All the activity is for a single grade denoted by *g*. The ovals on the left represent the inventories, with each oval signifying a different set of expertise. The ovals on the right indicate a set of billets at grade *g*, and their requirements, usually 1.0. Equation (1) relates flows of promotions to establish inventories of various areas of expertise. The arrows emanating from the inventory ovals and converging on the requirements ovals on the right are the assignments of inventory to billets. According to Equation (2), the total flow of assignments out of an inventory oval equals the inventory in question. Likewise, Equation (3) stipulates that the sum of assignments flowing into a requirement oval on the right equals the billet's requirement.

Figure E.1
Flows of Expertise to Billets

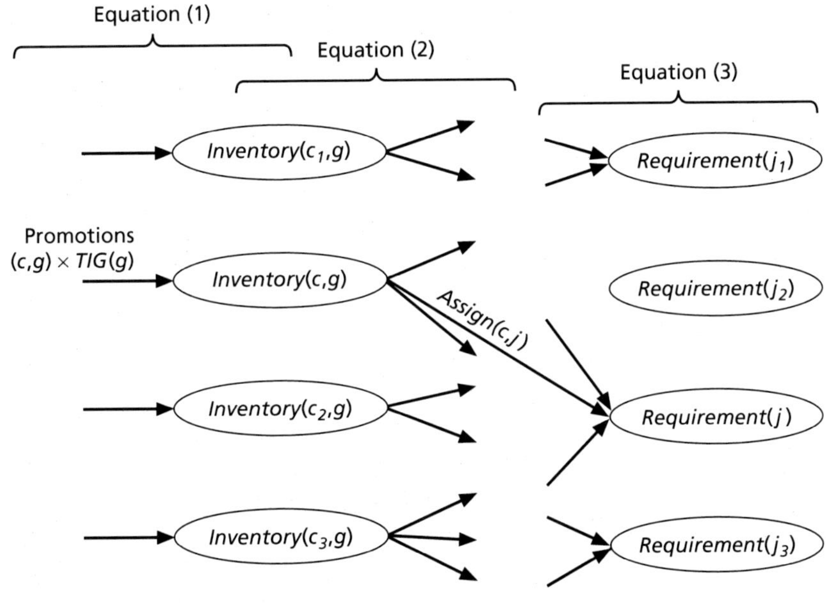

RAND *MG618-E.1*

Because we are dealing with steady-state averages, none of the variables we have introduced need to be integral, and fractional values are routine.

Bounds on Promotion Probabilities

The relationships discussed so far concern a single grade. Because flag officers are a closed system in that admirals enter as RDMLs and are promoted upward through grades or retire, there are relationships between grades that need to be observed. At the least, *Promotions(c,g)* has to be no greater than *Promotions(c,g* – 1) for any combination of areas of expertise. The flow between two successive grades is depicted in Figure E.2. The inventory that is promoted into grade *g* – 1 either is subsequently promoted into the next grade or retires without a promotion.

Figure E.2
Flows of Inventory Across Successive Grades

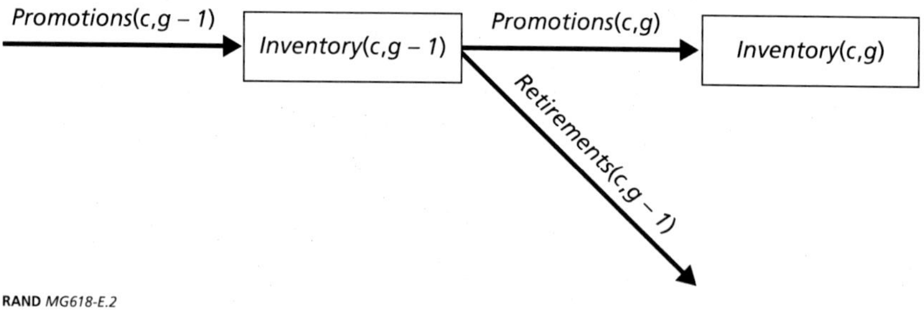

Equating the flows in and out of *Inventory(c,g* – 1) gives

$$Promotions(c,g-1) = Promotions(c,g) + Retirements(c,g-1),$$

and the probability of promotion is the ratio of the two promotion rates:

$$PromotionProbability(c,g) = \frac{Promotions(c,g)}{Promotions(c,g-1)}.$$

We require that promotion probabilities be no greater than one, but we can have greater control over solutions by constraining the promotion probabilities with upper bounds. This tends to produce solutions with broader ranges of competencies, and it can be used to shape the distribution of areas of expertise within the flag officer force. We specify upper bounds to individual promotion probabilities by expertise and grade, *PromProbUB(c,g)*. The bounds are enforced by the inequality constraints

$$\frac{Promotions(c,g)}{Promotions(c,g-1)} \leq PromProbUB(c,g). \tag{4}$$

The total promotion rate into grade *g* aggregated over all combinations of areas of expertise is fixed by the number of billets and average time in grade because of Little's Theorem. Moreover, some expertise-specific promotion rates are zero because there are no roles for the areas of expertise at the higher grade. In Chapter Five, we indicated a connection between these promotion probabilities and a notion of selectivity.

Thresholds on Promotion Rates

In Chapter Five, we noted that it is desirable to obtain solutions that do not have small positive values of promotion variables. These can be eliminated by setting thresholds such that promotion rates are either zero or at least equal to specified values. This can be accomplished, but the problem is now a mixed integer program rather than a simpler linear program.

We define *LowLimit(g)* to be the smallest value, other than zero, that we wish to allow *Promotions(c,g)* to be. We introduce "binary" variables, *Z(c,g)*, that are allowed to take on values of zero or one. For each expertise combination and grade, we include two inequalities:

$$Promotions(c,g) \leq M \times Z(c,g) \quad (5)$$

$$Promotions(c,g) \geq LowLimit(g) \times Z(c,g), \quad (6)$$

where M in Equation (5) is a number larger than any possible value that a promotion variable could take on. These work together in the following way. Having both *Promotions(c,g)* and *Z(c,g)* equal to zero is compatible with both inequalities. And, if one of the variables in a *(c,g)* pair is zero, the other must also be zero. If *Promotions(c,g)* is greater than zero, Equation (5) requires that *Z(c,g)* be equal to one, and then Equation (6) forces *Promotions(c,g)* to be at least equal to *LowLimit(g)*.

An Objective Function

We add an objective function and employ optimization software to find solutions that minimize it. The terms in the objective function involve the *Assign* variables corresponding to assignments of inventory to billets. For each possible kind of inventory, *c*, and job, *j*, there is a penalty "cost," *MatchCost(c,j)*, where smaller values are associated with better matches. The total "cost" of a solution is the sum of the assign variables weighted by their match costs:

$$TotalCost = \sum_{c,j} MatchCost(c,j) \times Assign(c,j). \quad (7)$$

Because we assume that any acceptable combination of primary and secondary areas of expertise is as good as any other for filling a billet, most values of *MatchCost(c,j)* are set to 1.0, and optimizing the objective function does not play a major role. But the objective function does have two purposes. As discussed in Chapter Five, there are billets for which any area of expertise, sometimes primary, sometimes secondary, and sometimes both, is acceptable. In such cases, it is preferred to have inventory with "any primary" and "any secondary" areas of expertise. To avoid naming specific kinds of expertise when the problem could be solved with an "any," we penalize the use of specific areas of expertise when an "any" would be feasible. There are occasions when such billets need to be filled with inventory with specific expertise in order to achieve a feasible solution.

The second role of the objective function is to aid in diagnostics. Sometimes in exploring boundaries, we introduce conditions such that there is no solution that satisfies the set of equations and inequalities. Normally when that happens the solver software would simply declare infeasibility and stop. To get more information, we added a set of extra areas of expertise that could satisfy any billet, but inventory involving these are heavily penalized by setting the corresponding *MatchCost* equal to 1,000. When inventory using these special areas of expertise has to be assigned to billets for mathematical feasibility, the solver terminates normally and we get a complete set of outputs that can be examined. Observing the *Assign* variables with the high penalties provides useful clues to where the problems lie. Without this trick, the solver would simply give up and we would not have any diagnostic information.

The Complete Model Formulation

Below is a summary of the formulation all in one place.

The variables are

- *Promotions*(*c*,*g*)
- *Inventory*(*c*,*g*,*j*)
- *Assign*(*c*,*j*)
- *Z*(*c*,*g*) (0 or 1)
- *TotalCost*,

where *c* is a set of areas of expertise, *g* is grade O-7 through O-10, and *j* is an index over the billets.

The constants are

- *Requirement*(*j*)
- *TIG*(*g*)

- *PromProbUB(c,g)*
- *M,* a number greater than any *Promotions(c,g)*
- *MatchCost(c,g,j)*.

The equations are

$$\text{Minimize } TotalCost = \sum_{c,j} MatchCost(c,j) \times Assign(c,j)$$

subject to

$$Inventory(c,g) = Promotions(c,g) \times TIG(g)$$

$$Inventory(c,g) = \sum_{j \in J(g)} Assign(c,j)$$

$$\sum_{c,j} Assign(c,j) = Requirement(j)$$

$$Promotions(c,g) \leq PromProbUB(c,g) \times Promotions(c,g-1)$$

$$Promotions(c,g) \leq M \times Z(c,g)$$

$$Promotions(c,g) \geq LowLimit(g) \times Z(c,g) \; .$$

References

Agence France-Presse, "US Is Planning to Station Second Carrier in Pacific," *Taipei Times*, August 12, 2004, p. 1.

Bigelow, Bruce V., "A New Take on the Warship," *San Diego Union Tribune*, February 5, 2005, pp. C1, C4.

Boraz, Steven C., "Behind the Curve in Culture-Centric Skills," *Proceedings*, Vol. 131, No. 6, June 2005, pp. 41–48.

Burgess, Richard R., "New Command Will Unify Navy's Maritime Force Protection Units," *Seapower*, Vol. 47, No. 9, September 2004.

Burke, Bill, "Navy Assembles a Tougher Defense Strategy as a Growing Number of Hackers Plague Its Front Line," *Virginian-Pilot*, June 26, 2005, p. A1.

Clark, Vernon E., USN, "Sea Power 21: Projecting Decisive Joint Capabilities," *Proceedings*, Vol. 128, No. 10, October 2002, pp. 32–41. As of October 23, 2007: http://www.navy.mil/navydata/cno/proceedings.html

———, remarks to Senior Executive Seminar, Falls Church, Va., September 4, 2003.

Fleishman E. A., L. I. Wetrogen, C. E. Uhlman, and J. C. Marshall-Mies, "Abilities," in N. G. Peterson, M. D. Mumford, W. C. Borman, P. R. Jeanneret, and E. A. Fleishman, eds., *Development of Prototype Occupational Information Network Content Model, Vol. 1*, Salt Lake City, Utah: Utah Department of Employment Security, Contract No. 94-542, 1995, pp. 10.1–10.39.

Gabarro, John J., *The Dynamics of Taking Charge*, Boston, Mass.: Harvard Business School Press, 1987.

Gayl, Franz J., USMC (Ret.), "Time for a Military Space Service," *Proceedings*, Vol. 130, No. 7, July 2004, pp. 43–45.

GlobalSecurity.org, "Fleet Response Plan," undated Web page. As of October 12, 2007:
http://www.globalsecurity.org/military/ops/frp.htm

Gugliotta, Guy, "Radical Warship Takes Shape: Navy's DD(X) May Launch Generation of Greater Efficiency," *Washington Post*, February 8, 2004, p. A03.

Harrell, Margaret C., Harry J. Thie, Peter Schirmer, and Kevin Brancato, *Aligning the Stars: Improvements to General and Flag Officer Management*, Santa Monica, Calif.: RAND Corporation, MR-1712-OSD, 2004. As of October 12, 2007: http://www.rand.org/pubs/monograph_reports/MR1712/

Hines, Jason, "Restore the Foundation of Naval Intelligence," *Proceedings*, Vol. 131, No. 2, February 2005, pp. 36–40.

Kilcline, Thomas J., USN, Director, Air Warfare (N78), "Helicopters at the Core of Strike Group Capability," *The Hook*, Winter 2004.

Kohart, Mark, "Protect the Force: Create a Navy Security Force Officer Corps," *Proceedings*, Vol. 131, No. 8, August 2005, pp. 75–77.

Koprucu, Feza S., USN, "Change How Information Is Managed and Controlled," *Proceedings*, Vol. 130, No. 8, August 2004, pp. 56–60.

Kreisher, Otto, "Service Experts Eye 'Leap Ahead' in Mine Warfare Capabilities," *Seapower*, Vol. 47, No. 9, September 2004, pp. 16–17.

LaFleur, Timothy, USN, "LCS Parts from the Past," *Proceedings*, Vol. 130, No. 7, July 2004, pp. 32–37.

Lawler, Edward E., "From Job-Based to Competency-Based Organizations," *Journal of Organizational Behavior*, Vol. 15, No. 1, January 1994, pp. 3–15.

Little, John D. C., "A Proof of the Queuing Formula: $L = \lambda W$," *Operations Research,* Vol. 9, May 1961.

McCall, Morgan, Jr., *High Flyers: Developing the Next Generation of Leaders*, Boston, Mass.: Harvard Business School Press, 1998.

Merle, Renae, "Navy Names Coastline Ship Contractors," *Washington Post*, May 28, 2004, p. E03.

———, "Navy Plans to Buy Fewer Ships," *Washington Post*, September 7, 2004, p. E01.

Morgan, Dan, "Proposed Ship Speeds into Gathering Storm," *Washington Post*, July 6, 2003, p. 5.

OPM—*see* U.S. Office of Personnel Management.

Patch, John P., "Sea Basing: Chasing the Dream," *Proceedings*, Vol. 131, No. 5, May 2005, pp. 38–43.

Poore, Ralph E., USN (Ret.), "The Value of Electronic Warfare Endures," *Proceedings*, Vol. 130, No. 9, August 2004, p. 37.

Prahalad, C. K., and Gary Hamel, "The Core Competence of the Corporation," *Harvard Business Review*, Vol. 68, No. 3, May–June 1990, pp. 79–93.

Reddy, Anitha, "Navy Contract Hurts EDS Earnings; Firm Takes $559 Million Charge," *Washington Post*, February 6, 2004, p. E05.

Robbert, Albert A., Steve Drezner, John E. Boon, Jr., Lawrence M. Hanser, S. Craig Moore, Lynn M. Scott, and Herbert J. Shukiar, *Integrated Planning for the Air Force Senior Leader Workforce: Background and Methods,* Santa Monica, Calif.: RAND Corporation, TR-175-AF, 2004. As of October 12, 2007: http://www.rand.org/pubs/technical_reports/TR175/

Schmitt, Eric, "Pentagon Said to Offer Cuts in the Billions," *New York Times*, December 30, 2004, p. 1.

Scott, Lynn M., Steve Drezner, Rachel Rue, and Jesse Reyes, *Compensating for Incomplete Domain Knowledge*, Santa Monica, Calif.: RAND Corporation, DB-517-AF, 2007. As of October 12, 2007: http://www.rand.org/pubs/documented_briefings/DB517/

Shippmann, Jeffrey S., Ronald A. Ash, Mariangela Battista, Linda Carr, Lorraine D. Eyde, Beryl Hesketh, Jerry Kehoe, Kenneth Pearlman, Erich P. Prein, and Juan I. Sanchez, "The Practice of Competency Modeling," *Personnel Psychology*, Vol. 53, No. 3, September 2000, pp. 703–740.

U.S. Naval Register, undated Web site. As of October 4, 2007: http://buperscd.technology.navy.mil/bup_updt/upd_CD/BUPERS/Register/Naval_Register.html

U.S. Navy, "CNO Guidance for 2005," January 3, 2005. As of October 12, 2007: http://www.cnsronline.org/archives/clark-guidance2005.pdf

———, *Manual of Navy Officer Manpower and Personnel Classifications, Volume I,* NAVPERS 15839I, October 2006.

———, Status of the Navy, Web page, September 28, 2007. As of October 4, 2007: http://www.navy.mil/navydata/navy_legacy.asp?id=146

———, United States Navy Biographies, undated Web page. As of October 4, 2007: http://www.navy.mil/navydata/bios/bio_list.asp

U.S. Office of Personnel Management, OPM Leadership Competencies, undated Web page. As of October 4, 2007: http://www.dtc.dla.mil/wfd/ldrshpdv/1.htm

Weisman, Jonathan, and Renae Merle, "Pentagon Scales Back Arms Plans: Current Needs Outweigh Advances in Technology," *Washington Post*, January 5, 2005, p. A01.

Wheeler, Warren C., "Naval Intel Must (Re)Acquire Core Skills," *Proceedings*, Vol. 130, No. 2, February 2004, pp. 42–47.

Winand, Timothy E., "Space-Based Weapons," *Proceedings*, Vol. 130, No. 7, July 2004, pp. 43–48.